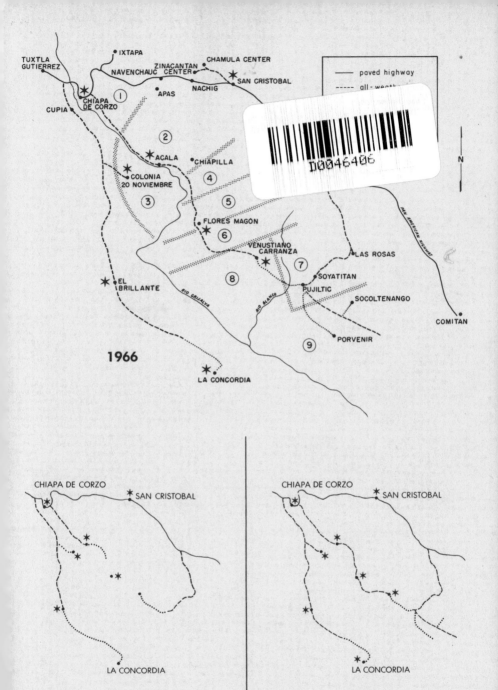

Change and Uncertainty in a
Peasant Economy

Change and Uncertainty in a Peasant Economy

The Maya Corn Farmers of Zinacantan

Frank Cancian

Stanford University Press, Stanford, California   1972

Stanford University Press
Stanford, California
© 1972 by the Board of Trustees of the
Leland Stanford Junior University
Printed in the United States of America
L.C. 72-153814
ISBN 0-8047-0787-1

To Maria and Steven

Acknowledgments

MOST OF the fieldwork for this study was done in 1966–67 with the support of a postdoctoral fellowship in Latin American Studies from the Foreign Area Fellowship Program. The fellowship also gave me free time before and after the fieldwork. Small grants from the Latin American Studies Program at Cornell University and the Latin American Studies Program (in cooperation with the Center for Research in International Studies) at Stanford University paid for assistants during analysis, and a grant from the Morrison Fund of Stanford University permitted me to make a trip to Mexico in the summer of 1965.

Progress on the manuscript was twice delayed by world and campus troubles, at Cornell in 1969 and at Stanford in 1970, and the final draft was completed at the Center for Advanced Study in the Behavioral Sciences (Stanford) in 1970–71. I am indebted to the Center for the atmosphere it provides, and to the National Institute of Mental Health Special Fellowship (MH17719) which permitted me to enjoy it.

My greatest debt is to the hundreds of Zinacantecos who extended courtesies and permitted the exchange of favors. José Hernandez Perez, José Hernandez Nuh, Guilliermo Perez Nuh, Manuel Perez, and Mariano Martinez were creative assistants as well as valuable informants—as was Teódulo Martinez Osuna, who lives in San Cristóbal.

When I went to the lowlands, I depended on the Zinacantecos

and Ladinos who were my hosts. Shun Vaskis of Navenchauc, Shun Vaskis of Pih, Domingo de la Torre Perez, the late Pedro Perez con Dios, and Guilliermo Perez Nuh shared their camps and their knowledge with me. Among the many landowners who were helpful, I want especially to thank Jorge Osuna and Hernan Pedrero. Life in San Cristóbal was enriched by anthropologists connected with the Harvard Chiapas Project and by local residents. It is a pleasure to thank George and Jane Collier, who were my neighbors and coworkers for the entire field period, and Evon Z. Vogt, director of the Harvard Chiapas Project. Graciela Alvarado vda. de Villatoro and Gustavo Hernandez S. granted small but important favors.

Muriel Torrey, Susan Almy, and Pamela Oliver were valuable research assistants. George Collier, Stuart Plattner, Renato Rosaldo, Evon Vogt, and Pan Yotopoulos read drafts of the manuscript and fostered many improvements. Before, during, and since the fieldwork my wife, Francesca, has provided professional counsel, help, encouragement, and lots of fun.

F. C.

Contents

Tables

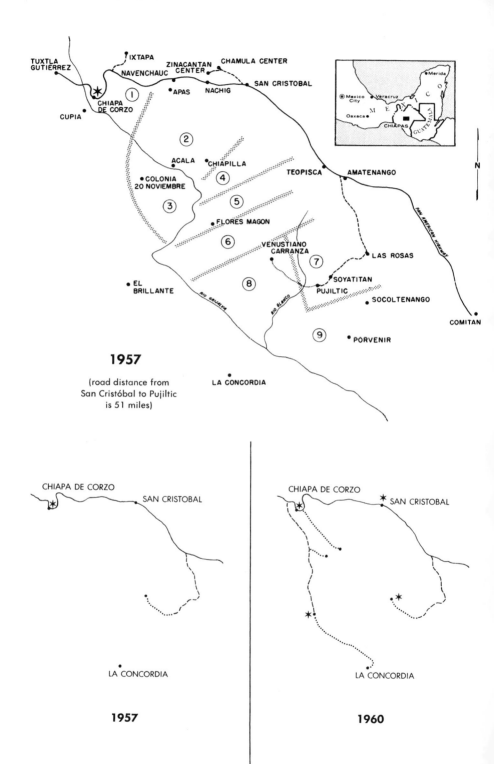

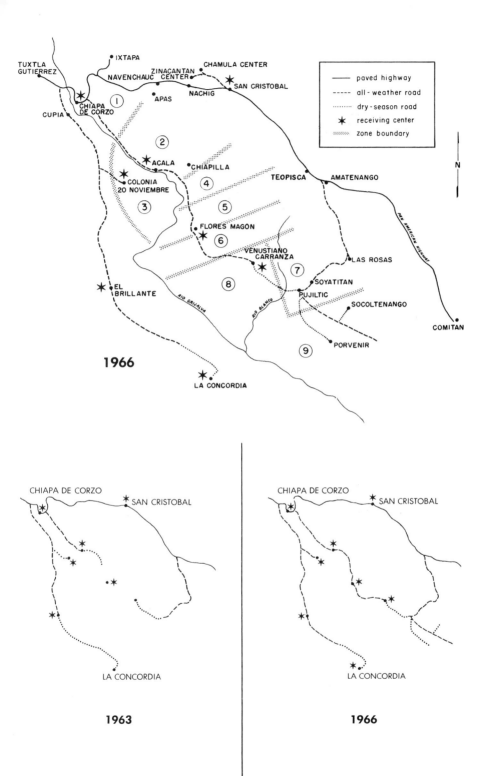

Change and Uncertainty in a
Peasant Economy

1. Zinacantecos in the lowlands return to their temporary shelters
after a day of weeding in the fields. The corn leaves are curled
after a long dry spell.

ONE

Introduction

THIS IS A STUDY of economic change in Zinacantan, a township of some 9,000 Maya Indians in the highlands of Chiapas, Mexico. Zinacantecos are corn (maize) farmers who work with hoes and the techniques of slash and burn agriculture to produce the major part of their families' food, as well as substantial amounts that they market for cash. Most of this corn farming is done on rented fields in the nearby lowlands. The empirical focus of this book is Zinacanteco farming and the changes in it resulting from government development programs between 1957 and 1966. The theoretical focus is on uncertainty—on how lack of knowledge creates situations in which even economic maximizers, rational economic men, have their economic decisions substantially influenced by noneconomic factors. In this Introduction I will briefly outline both aspects of the study.

Any study of peasant agriculture done today is faced with the perennial "economic man" issue, as it was stated in the literature on agricultural development published during the 1950's and 1960's. The classic formulation of the issue would ask whether Zinacantecos are economic maximizers or are kept from maximizing by their adherence to tradition. That is, the economic rationality of Zinacanteco responses to changing economic opportunities would be in question. In this study I have avoided this issue as much as possible, because: (1) I believe that it is a bogus question and leads to scientifically incorrect and politically dangerous descriptions of peasant societies; (2) I want to focus on the much more limited uncertainty question and must view Zinacantecos as economic maximizers in order to deal with this question.[1]

An Outsider's View of Change

An oversimplified outsider's version of what has happened might go something like this. In 1956 the Zinacantecos were farming corn on rented lands in the Grijalva River Valley below their homes in the highlands. The area they worked was all within a day's walk of their homes, and at harvest time the corn was packed up to the mountains on muleback. Since most farmers harvested considerably more than was needed for household consumption, they sold the excess in the nearby city of San Cristóbal, usually to local Ladinos (Spanish-speaking non-Indians) and to Indians from corn-poor communities in the highlands. The most economically successful Zinacanteco farmers were those who could withhold their corn from the market until prices reached a peak in August.

Then various branches of the Mexican government began two important programs. A major all-weather road was cut through the Grijalva Valley and numerous feeder roads were built in both the lowlands and the highlands (see my maps, pp. xii–xiii). Zinacantecos could now reach fields that had been too distant in the days of mule transport, and some of these fields were substantially more productive than those within walking distance of Zinacantan. The second program established government receiving centers and warehouses for corn near both the old and the new fields. These centers bought corn in wholesale lots and paid a fixed price throughout the year. From the farmers' point of view, the government price usually compared favorably to the San Cristóbal price, especially since the cost of transport from the cornfields to

---

[1] The classic formulation of the economic man issue often leads to implicit comparisons in which peasant customs that "impede" economic change are described as if they existed without parallel in modern economic systems. That is, the cultural context of peasant economic systems is examined in a way suggesting that our own economic system functions free of cultural constraints. But in fact, all economic systems are substantially embedded in a cultural or institutional matrix. This point and the problems it creates for those concerned with meaningful comparative research and the formulation of development policy are discussed in Appendix D, pp. 189–99.

the receiving centers was much lower than the cost of transport to the highlands. Moreover, the farmer could sell in wholesale lots in November or December rather than dealing with uncertain market fluctuations the next summer.

In general, an outsider looking at the situation would have advised Zinacantecos to abandon mule transport and retail marketing in favor of motor transport and wholesaling to the new receiving centers. In fact, this is just what some Zinacantecos did. With each passing year, more and more of them moved their farming operations to the more distant fields, shifted from mule to truck transport, and sold much of their crop to the receiving centers.

Change from the Zinacanteco Point of View

An outside observer comes to the situation with two advantages: the benefits of hindsight, and an income that is not directly tied to the accuracy of his analysis. Since the Zinacanteco farmer had neither of these, and since his sources of information were limited, a different picture of the situation must be constructed for him.[2]

In 1956 the farmer could walk to fields near his village, taking his crop home with mules, and retailing part of it in the San Cristóbal market. He might follow the same routine year after year, or he might change fields and farming techniques every year in search of the most profitable combination; but in any case his alternatives were limited. More important, he had a fairly clear idea of the results to be expected from a given course of action, for the system within which he raised and sold his crop, despite minor variations, had existed for many years.

When the government programs began, roads opened and trucks became available. A Zinacanteco might have found truck transport very desirable, especially if he normally had few mules or had just lost mules and faced a major investment for new ones. Truck transport had a fixed price that could be determined on the spot, and it did not involve the physical discomfort of mule transport. The

[2] The point of view I have attributed to Zinacantecos is not intended to be a "culturally specific" point of view based on systematic investigation of native categories of thought. Rather, it is my interpretation, based on many conversations with Zinacantecos, of what it might have been like to live through the period under discussion.

farmer might have been reluctant to give up his mules, since they served both as status symbols and as sources of extra income when he was not farming corn. And if he customarily rented mules, he might have been reluctant to abandon his working relationship with the mule owner. But the costs and benefits of truck transport were readily apparent.

The other option offered by the new roads was not so simple. Was the farmer to work fields in the new areas opened by the roads? He had heard reports, usually secondhand, that the new fields were fantastically productive. However, there were many uncertainties involved in a move. His corn was adapted to micro-environments that differed from place to place even within the old area, so he would have had to obtain seed from someone who was already working the new area. His contacts among the Ladino residents and Zinacantecos in the old area would no longer have been able to provide seed corn, or the other favors (such as places to sleep and store tools) that were so much a part of work in the lowlands. Moreover, there was no getting around the fact that one had to pay for transport to the new area, since it was too far to walk; so a large cash investment in carfare would be lost if the crop failed. And he had heard that the rains came at a different time in the new area. How could he be sure when to seed and when to return for the weeding? On the whole, there were many uncertainties, and many new problems that might counterbalance the benefits of the higher yields that the new fields were said to give.

Sale to the government centers involved somewhat similar uncertainties. People said the centers would pay well for corn, but was this rumor about light corn or heavy corn? Receiving centers weighed the corn, whereas most farmers were accustomed to selling by volume measures; and centers also discounted for dampness and impurities in the corn. To sell to the government, the Zinacanteco farmer also had to face the Spanish-speaking bureaucracy and concern himself with scales, receipts, and trips to banks. If the price of corn in San Cristóbal were especially high the following August, sale to a receiving center might not be so good, for the farmer would miss a chance to sell at the peak price.

In general, anyone accepting this view of the situation, especially in the early years, would have advised Zinacantecos to wait for more information. Yet many of them in fact decided to move their fields and sell to the receiving centers, even though the potential extra profits were very uncertain at best.

Economic Man, Uncertainty, and Innovation

Uncertainty is one of those elements of decision-making that are conventionally avoided by microeconomic theory. The "economic man" who resides at the core of microeconomic theory is a caricature self-consciously created by economists to permit development of their theory. He is based not on experience but on the need to make simplifying assumptions that will allow one to interpret behavior.[3] As Dorfman puts it:

> At bottom, an economy consists of people, and people are more complicated than we care to reason about. To keep things manageable, economists populate their theories with *Homo economus*, a very simple species of mankind who know a great deal, reason accurately, behave predictably, and are driven by uncomplicated motivations, quite unlike you and me. *Homo economus* always finds out when the price of bread rises and promptly consumes less of it. (1964: 11–12.)

Dorfman hastens to add: "In real life the consequences of economic decisions are never really known. . . . Uncertainty is an ineluctable fact of economic life" (1964: 12). Thus economic theory, based on economic man, is not intended to accurately describe the behavior of the actor who does not "know a great deal."

In the early chapters of this book I have presented Zinacantecos as economic men. Zinacanteco agriculture is described in terms of inputs and outputs that are relatively stable and calculable, and I have avoided reference to the increased uncertainty that was produced by change. This description is done *within* the framework of cultural conventions and institutional constraints common to Zinacanteco farmers. Customs concerning the treatment of workers, for example, are taken as part of the framework and are described in detail. Although I have considered the economic impli-

---

[3] Whether economic theory is normative or positive—i.e., prescriptive or descriptive—is a question many practicing economists seem to have answered by simple inattention. For now, I will follow their lead.

cations of hiring few or many workers, given the customs, I have not analyzed the "efficiency" of the customs in economic terms. With this view of Zinacantecos as economic men it becomes possible to appreciate the importance of uncertainty to their behavior in the face of change.

The book argues that because of the uncertainty about outcomes of new activities, the economic man model cannot provide a complete framework for description. In later chapters I will try to show that the difference between Zinacantecos who took up the new practices relatively quickly and those who did not can be explained —in part—by a general theory of differential response to uncertainty. The empirical success of the theory and the intuitive appeal of "subjective" descriptions like that presented above are jointly reinforcing arguments for the importance in Zinacanteco economic life of decisions made under uncertainty.

Although I am not competent to place this effort within the context of the study of risk and uncertainty by economists, I do want to record some impressions based on limited experience with the literature. Knight's classic distinction between risk and uncertainty, to which I subscribe, seems to have been overwhelmed by the demands of the microeconomic model for a characterization of decision-making in calculable terms. Knight (1921: 19–20) distinguishes between measurable or quantifiable uncertainty, which he calls "risk," and unmeasurable or "nonquantitative" uncertainty, which he calls "uncertainty." Risk describes situations where an actor knows the odds for and against a desirable outcome from a given course of action; uncertainty describes situations where he does not. Economists seem to translate uncertainty into terms that make it comparable with risk. In many recent discussions of agricultural development, "risk and uncertainty" appears as a single term, not as a conjunction of concepts denoting differentiable phenomena.[4] Thus the microeconomic framework for analysis be-

---

[4] Wharton (1968), who is very sensitive to the importance of Knight's distinction for the interpretation of peasant economic behavior, stresses the actor's point of view and reconceptualizes the uncertainty problem in terms of subjective probability (1968: 23ff). The essence of subjective probability, as he uses the term, is that it cannot be determined in terms of the objective factors normally considered by the economic analyst.

comes untenable when uncertainty is an important factor—except in prescriptive (normative) economics, where much has been done to derive optimal procedures for decision-making in uncertain situations. When descriptive (positive) economics needs to include uncertainty in its model of decision-making, it is forced to examine the noneconomic characteristics of actors.

The nature of my argument, and the limits of its applicability, may be illuminated by contrasting it with two well-known traditions of research in the study of agricultural development: diffusion of innovations, and achievement motivation. The literature in rural sociology on the diffusion of innovations is essentially consistent with the applicability of microeconomic theory to situations of change. This literature (see Rogers 1962) tends to focus on the diffusion of information and explains innovative behavior in terms of the availability of information. It is similar to my approach in that it focuses on the differential response of individuals to new opportunities; and it provides an explanation for the part of differential response that is not affected by the uncertainty factor.

My position is also similar to that of many students of achievement motivation (McClelland 1961, Atkinson and Feather 1966) in that it deals with aspects of innovative behavior that cannot easily be fitted into the microeconomic framework, given its focus on informed decision-making.[5] The difference between my approach and that of the psychological students of achievement motivation lies in the independent variables used to predict an actor's willingness to take a chance in uncertain situations. Some studies by psychologists do relate achievement motivation to social characteristics, but most are ultimately concerned with the psychological variable as it is measured by projective tests and approximations to projective tests. In contrast, I am ultimately concerned with rank in a stratification system as the independent variable.

[5] If one uses concepts like subjective probability, these conceptualizations can always be rephrased in terms of a microeconomic decision-making framework with a "strong" information assumption. This is a logical property of the maximization framework, not necessarily a characteristic of human decision-making. The fact that individuals are viewed as treating the "same" information differently according to their achievement motivation or their rank is an important departure from the objective viewpoint of economics.

In sum, my purpose is to illustrate the importance of the uncertainty that characterizes change situations and to suggest the limitations on the usefulness of the microeconomic theory that result from the uncertainty factor (especially in the early stages of change, when information is in short supply and uncertainty may be seen as a major component in decision-making). On the positive side, I hope my theory relating innovative economic behavior to economic rank will contribute to an understanding of the social structural factors that influence economic behavior.

Plan of the Book

Chapter 2 gives a general description of Zinacantan and a brief overview of the township's economic life. After that, the book falls into three parts. The first (Chapters 3–6) is an ethnographic description of Zinacanteco corn farming as of 1966. Chapter 3 describes the organization of agricultural production and gives basic facts about Zinacanteco farmers, the landowners from whom they rent their fields, and the workers they employ to help them. Chapter 4 describes the agricultural work cycle, including details on labor inputs and yields under the various conditions Zinacantecos encounter in their farming; and Chapter 5 describes what is done with corn after the harvest, i.e., the transportation and marketing system. The extensive detail in Chapters 3–5 can be justified on the ground of its ethnographic usefulness. In addition, I have used this information to support my summary description of the costs and yields of farming in terms of standard units. In Chapter 6 these standard units are used to estimate the net return or profit to farmers who work in many different ways. The reader who is more interested in economic change than in the details of the economic system can get the necessary background for my discussion of change by reading the introductory section and the first and last paragraphs of the other sections in Chapter 3, together with the first and last sections of Chapters 4, 5, and 6. I hope that the occasional repetition necessitated by my attempt to serve those not interested in ethnographic detail will not burden those who are.

Chapter 7, the second part of the book, focuses on the changes that took place during the decade 1957–66. Here, the description and analysis is done from the aggregate rather than the individual point of view. That is, I am not concerned with the response of particular types of Zinacantecos to the economic opportunities offered by the government's programs, but with the response of Zinacantecos in general. The effects of the programs are highlighted by a brief comparison of two hamlets that differ in their access to roads; but the chapter as a whole is an attempt to show that many Zinacantecos have responded with apparent rapidity to the opportunities offered by the government programs. How fast or slow the responses or changes have been is a comparative question that I do not attempt to answer; but the chapter suggests that characterizing Zinacantan as a stagnant peasant community would be silly.

Chapter 8, the final part of the book, develops and tests a theory relating behavior under uncertainty to position in a stratification system. Data on Zinacanteco responses to the government programs described in earlier chapters are used to test the theory, and data from a previous comparative study (Cancian 1967) are displayed as additional support for the theory. This chapter provides evidence for the importance of uncertainty in change situations, and for the importance of social structural variables in predicting economic behavior under these conditions.

Types of Data

Much of the information presented below was gathered with traditional anthropological research techniques: open-ended interviewing, "participant observation," and the collection of documents, especially maps. Since many of the questions I wanted to ask are best dealt with statistically, I also gathered and used two other types of data. "Work histories" are systematic interviews on the details of corn farming as practiced by twenty individuals during the years they worked independently as adults. For these interviews representativeness was sacrificed so as to increase the detail and reliability that could be achieved only with intelligent, motivated informants who knew me well enough to be fairly open. The "surveys" include virtually all the adult males (about 300) from two hamlets. These men were interviewed about less sensitive and less detailed matters by Zinacanteco assistants. Tables in the text indicate which (if either) of these sources of data they represent, and Appendix A describes more fully the procedures used in gathering these and other data. The reader who wants to explore the methodology at the outset should probably read Chapter 2 before reading Appendix A.

Economic Life in Zinacantan

THIS CHAPTER will briefly describe the context in which corn farming exists. The few facts I present will give the reader background for the chapters that follow; but they cannot evoke the many rich aspects of Zinacanteco life that are not discussed in this study. Much of what anthropologists know about Zinacantan has been published in recent years, by members of the Harvard Chiapas Project. Evon Vogt's comprehensive ethnography *Zinacantan: A Maya community in the highlands of Chiapas* (1969) is by far the best source of information on the Zinacanteco way of life as a whole; and the reader who wants a quick introduction will find it in Vogt's *The Zinacantecos of Mexico: A modern Maya way of life* (1970). The existence of these books, and many other works by Vogt's students and colleagues (see Vogt 1969: 703–22) allows me to limit the focus of the present study. Finally, I hope the photographs scattered throughout the book will give the reader a feeling for the setting in which this study was done.

Zinacantan

Zinacantan is a township populated by about 9,000 Tzotzil-speaking Maya Indians and a handful of Spanish-speaking Ladinos. It lies on both sides of the Pan-American Highway just west of the city of San Cristóbal de Las Casas (population 23,000) in the highlands of the state of Chiapas. The township (*municipio*) is a political subdivision introduced by the Spanish colonial government, but it coincides almost exactly with the clearly defined linguistic and cultural community of the Zinacantecos.

Besides the use of Mayan languages, the characteristic that most clearly distinguishes the Indians of the region from the Ladinos is dress. The more than 150,000 highland Indians who use San Cristóbal as a market center wear about a dozen distinctive non-European costumes, each peculiar to a different community. The groups are endogamous, and there is a tendency toward economic specialization by township. For example, the closest neighbors of the Zinacantecos are the Chamulas (Pozas 1959), who have similar, but readily distinguishable, language and customs; they wear a completely different type of clothing and concentrate less on corn farming and more on other economic activities. Other groups in the area are even more strongly distinguished from the Zinacantecos and from each other. In their own view, Zinacantecos are Zinacantecos first, Indians second, and Mexicans last, if at all.

The township is spread over 117 square kilometers of mountainous country. The ceremonial and political center, Hteklum, is 2,152 meters above sea level;[1] and the 15 outlying hamlets (*parajes*) are at varying altitudes within the cool, highland area. Density of settlement varies. In some hamlets groups of houses are scattered over oak-covered slopes. The people of Navenchauc live close together in a fertile valley where a small lake spreads to cover part of their cornfields during the summer rainy season. And in the hamlets to the west sloping, virtually treeless cornfields separate groups of houses by hundreds of yards. Hteklum lies in a valley at the northern edge of the township. There, streets and lanes form haphazard blocks around a central plaza, which is bordered by the principal church, the town hall, a school, the medical clinic of the Instituto Nacional Indigenista, and a dozen or more small Ladino and Indian shops.

---

[1] I have used the native word, Hteklum, for the ceremonial and political center of Zinacantan, and have used Spanish renderings of other hamlet names for the sake of consistency with my other publications and with the many Spanish place names outside Zinacantan to which I will refer. In many publications on Zinacantan, phonetic renderings of Tzotzil names are used; thus "Nachig" is /nachih/, "Apas" is /7apas/, and so on.

2. The hamlet of Nachig from the air. At the lower left the Pan-American Highway leaves the valley in the direction of San Cristóbal.

3. The hamlet of Apas from the air. The beginning of the drop to the lowlands shows in the upper part of the picture. (Both photos on this page were taken by the Cia. Mexicana Aerofoto, S.A., and are used by courtesy of the Harvard Chiapas Project.)

This study draws heavily on data from two hamlets. The larger, Nachig, has a population of about 1,000. It lies on either side of the Pan-American Highway a few miles west of San Cristóbal. For fiestas, Nachig people walk over the mountain that separates them from Hteklum, a journey of about two hours; but they can travel by truck to San Cristóbal in less than half an hour. The approximately 600 people of Apas, the other hamlet, are more isolated. To reach San Cristóbal they must first walk for an hour to the highway and then take a truck, which costs twice as much and takes twice as long as it does from Nachig. However, Apas stands virtually on the edge of the escarpment that falls to the Grijalva River Valley, so that its men have easy access to the fertile fields in that hot, lowland region.

Consumption

Zinacantecos purchase a small part of their food, clothing, and household goods outside the township, but they produce most necessities themselves. Although day-to-day consumption varies substantially from family to family, it is limited to a few standard items. (Vogt 1969 provides a detailed description of material culture and consumption.) The local diet is based on corn and beans, supplemented by squash, green vegetables, and fruit. Meat, which the "average" family eats perhaps once a week, may be purchased in San Cristóbal or may come from chickens and pigs raised by the household. Many families consume eggs from their own chickens more frequently than meat. In addition, Zinacantecos do purchase salt and a few luxury products such as coffee, sugar, liquor, beer, and soda pop; but the basic diet is overwhelmingly made up of items produced by members of the household.

Clothes are usually woven by Zinacanteco women from cotton thread purchased in San Cristóbal or wool produced locally. Virtually all women know how to weave the full range of clothing in the normal Zinacanteco costume; but a few items, like ceremonial clothing and the intricate blouses presented to godchildren, are often produced by specialists within the community. Men usually buy sandals and leather purses from artisans in San Cristóbal;

women do not wear shoes. Sheets of plastic for rainwear, as well as the trousers and shirts that men frequently wear for trips to the lowlands, are purchased outside the community.

Houses are typically one-room structures. Many Zinacantecos still build their own houses, using wattle-and-daub walls and straw roofs; but more and more are beginning to pay for the construction of adobe houses roofed with tile or corrugated metal purchased in the city. Major dams are now being built on the Grijalva River, but at the time of this study no electric power was available in the township. In 1966 Hteklum, the ceremonial and political center, had a recently built water system with pipes bringing water from the mountains; but other hamlets still depended on shallow communal wells.

Tools, including the standard machete (which is manufactured in Hartford, Connecticut), dishes, candles, and lanterns are purchased in San Cristóbal, as are soap and a few other household conveniences. A few Zinacantecos have radios or wristwatches, and many of the small stores in the township have installed record players and loudspeaker systems. Politicians, and occasionally others, employ Ladino lawyers when they are involved in disputes that cannot be settled in the community (J. Collier 1970). And almost all Zinacantecos have small regular expenses for local taxes and transportation to and from the San Cristóbal market.

Though many families buy medicines and occasionally pay clinic fees in San Cristóbal, the greatest expenditures for health care go for curing ceremonies performed by native shamans. Birth, death, and especially marriage also involve extraordinary expenses.

Most Zinacanteco ritual or ceremonial expenditures are made in the service of *cargos*, religious offices connected with the local form of Catholicism. Cargo posts are filled by the men of the community for one-year terms, with some men serving as many as four times (Cancian 1965a). Most families spend at least a few thousand pesos[2] in cargo service over the years, and much of the saving that Zinacantecos do has cargo service as a goal.

[2] One peso equals 8 cents U.S. In this study I have used the dollar sign ($) to indicate pesos, as is done in Mexico.

Zinacantecos require relatively small amounts of cash for the expenses of everyday living. My guess is that they spend substantially less than the value of the food and other items produced within the household. On the other hand, life crises and ceremonial expenditures, especially those in the cargo system, provide an outlet for extraordinary amounts of cash when it is available. In recent years, however, the expansion of the population and the economy has outstripped the economic demands of the cargo system. More and more Zinacantecos have purchased expensive products manufactured outside the community and the region (Cancian 1965a).

Supplementary Production Activities

Although lowland corn farming is the most important economic activity in Zinacantan, other activities occupy most men part of the time, and a few specialists do no farming at all. With very few exceptions, male economic activities fall into three categories: wage labor, trade, and farming.[3]

Many young men who are saving for the expenses of courtship and marriage (J. Collier 1968) leave the community to do road work, for relatively high wages are often available. This kind of work was especially popular in the 1950's, when the Pan-American Highway was being paved through Zinacantan, but there are still many jobs available on road and building construction in the area. It is safe to say, I think, that the men who were married by the 1960's (that is, those in their early twenties or older in 1970) still considered trade and farming more appropriate activities for adult men. A very few Zinacantecos depend on agricultural labor for their fellows as a principal means of livelihood; and, as will be shown in Chapter 3, many farmers occasionally do work of this kind.

Trade is a traditional Zinacanteco economic activity. Besides the retailing of corn and beans, which is a basic part of the farmer's

[3] The exceptions include truck drivers, full-time storekeepers, and employees of the Instituto Nacional Indigenista—in all, less than a dozen. A very few Zinacanteco men produce craft items on a part-time basis, and rare individuals occasionally produce illegal liquor.

role, Zinacantecos trade salt, fruit, and flowers. A number of families in Hteklum and nearby hamlets have traditionally been salt merchants for the highland region, transporting salt from Ixtapa, where it is made, and retailing it in San Cristóbal and the Indian communities of the highlands. And many farmers occupy spare time between farming obligations with salt selling. The bulking, transport, and retailing of fruit is another important sideline. Before the advent of the modern transportation system Zinacantecos who owned several mules were in a particularly good position to do this kind of trading, and also to do contract hauling of coffee from the area around San Cristóbal. The new roads have made mules less important to the region (see Plattner 1968), but many Zinacantecos have continued small-scale fruit trading, using the many trucks that carry small traders from place to place. The availability of motor transport along the Pan-American Highway to Tuxtla and San Cristóbal has also made it possible to produce fresh flowers for sale in the cities (Bunnin 1966). A good many farmers have a hand in the flower trade; but in 1966 there was only one man who tried to make it his full-time occupation. Finally, a few men began in the 1950's to concentrate their efforts on transporting and retailing corn produced by others (see Capriata 1965 and Chapter 5 below).

Although the production of flowers, fruit, and occasionally coffee has become more important of late, the vast majority of Zinacantecos still concentrate on raising corn and beans. Beans are planted in cornfields and are harvested separately from the corn. In the highlands beans are seeded with corn at the same time and in the same hole; in the lowlands they are seeded in late August or early September, when the corn is virtually mature. Lowland bean farming thus requires an additional weeding of the fields, which were first weeded in June and July to protect the corn itself from weed competition. Lowland fields vary greatly in their ability to support a bean crop interplanted with the corn, and some farmers who do not plant beans are prevented from doing so by their choice of cornfields. Occasionally, farmers working in one area of the lowlands "borrow" cornfields in another area and seed beans.

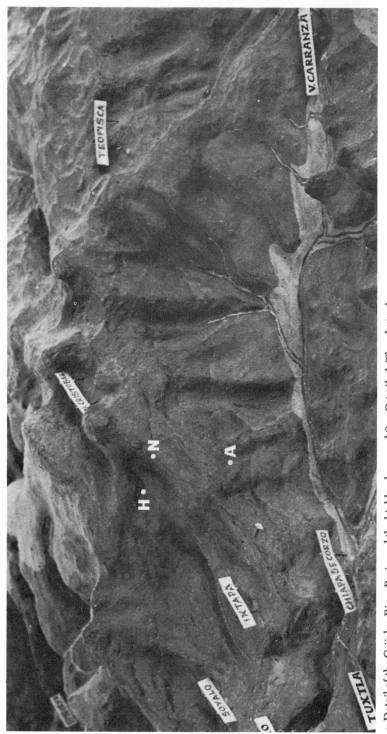

4. Detail of the Grijalva River Basin and the highlands around San Cristóbal. The photo shows part of a relief map of the state of Chiapas on display at the state botanical museum in Tuxtla Gutiérrez. The locations of Nachig, Apas, and Hteklum (the ceremonial and political center of Zinacantan) are marked by capital letters.

In sum, although bean crops provide a very substantial part of farming income for some Zinacantecos, substantial lowland bean production is infrequent enough that my data do not permit me to treat it systematically in this study. The cases in Appendix B show how the importance of beans varies from farmer to farmer.

Corn Farming

Lowland corn farming is often supplemented by work on fields elsewhere. Some men rent highland fields from other Zinacantecos, but fields in this area are typically owned by the farmer and located near his home. Many highland fields, and much of the land on the slopes leading to the Grijalva Valley, are held by Zinacantecos as *ejido* ceded to them under the Mexican land-reform program (Edel 1966, Vogt 1969). All the lowland fields worked by Zinacantecos are rented from non-Zinacantecos, and many of them are on large ranches that have survived the land reform.

In his study of land use in Apas, George Collier (1968) found that highland yields are typically lower; but that the many extra expenses (e.g., travel, rent, and so on) required for lowland farming make the overall return to labor and capital greater in the highlands. However, productive highland land is limited in quantity. Activity in Apas fields increased from the 1940's until about 1960 and then declined substantially, presumably because of the need to leave land fallow. Collier concludes (1968: 86):

During the periods of most intense land use the area may have provided over twice the current [1967] contribution to family support, but it never provided the average family with more than food. . . . It is obvious that Apas has had to rely on lowland farming to supplement production on its own land since 1940. While the land may have yielded a significant proportion of family income from 1945 to 1960, more recent farming has turned markedly to rental of lowland fields, and the land area is dwindling in its economic significance.

My own figures for 1966 show that 24 percent of Apas men and 37 percent of Nachig men were growing corn in highland plots, whereas more than 90 percent in each hamlet had lowland fields. However, less than 10 percent of the men in each hamlet planted enough highland corn to make their efforts a major economic activity; and more than 75 percent have major fields in the lowlands.[4] Some of the hamlets near the lowlands may be presently enjoying a peak in the production of their highland fields similar to that described by Collier for Apas in 1960; and some may even be harvesting the major part of their crop from such lands (see Price 1968). It is clear, however, that the lowlands are the principal site of corn production by most Zinacantecos.

The Grijalva Valley itself is extensive, and climate varies somewhat within its area (Helbig 1964). The mountainous terrain surrounding the valley, the many small tributaries of the Grijalva, and the rudimentary road system all make travel to and from the fields a major part of the Zinacanteco farmer's work. Whether they pay for passage on a truck or walk down the escarpment, farmers must spend the better part of a day in reaching their fields. Hence they go for periods of a few days to two or three weeks. They usually leave their families at home, live in temporary shelters, and eat staples they bring with them from the highlands.

Since distance and transport are important elements in the economic picture, I have divided the Grijalva Valley into the nine zones that are indicated on the maps (see pp. xii–xiii). It will become clear that adjacent zones often share characteristics. The division into zones is based on the time it takes to walk from Nachig and Apas to the various estates or ranches within the zones (see Tables 5.1 and 5.2).

If the reader will study Photo 4 and remember that hundreds of Zinacantecos leave their homes in the cool highlands, descend the escarpment, and spend days and weeks farming rented land spread across the extent of the hot Grijalva Valley, he will have the general background necessary to follow the detailed exposition in succeeding chapters. A glance at the photographs of farming throughout the book should give him a feeling for the terrain and the technology.

[4] Details on the size of lowland farming operations by Apas and Nachig men will be given in later chapters. Here, the higher percentages (90, 24, and 37) refer to farmers with one or more almuds seeded and the lower percentages (75 and 10) refer to farmers with two or more almuds seeded. (Local weights and measures are discussed in Appendix C.)

THREE

The Organization of Production

VIRTUALLY EVERY Zinacanteco man is a farmer. In organizing his work in the lowlands he must rent land; and in most cases he must hire workers to help him during the peak periods of the agricultural cycle. Four other roles are important in these activities: group leader, employer, worker, and landowner. The farmer himself may fill one or more of the first three roles, but Zinacantecos are never landowners. The eight sections of this chapter will present a detailed picture of the organization of production in the lowlands and the constraints within which a farmer normally works. The reader who wants a concise summary of production may get it by reading only the first and last paragraphs of each section.

Age, Wealth, and Farming Roles

Though all farmers work lowland fields for their own profit and pay rent on an individual basis, not all farmers deal directly with the landowners. Usually, a group of farmers will deal with a landowner through one of their number, the group leader. Thus all group leaders are farmers, but relatively few farmers are group leaders. Most farmers employ workers at one time or another. A few Zinacantecos who raise very little in their own fields supplement their incomes by working for other farmers, but most hired workers are Indians from other highland municipios, especially Chamula. Although a given farmer may take all the other roles except that of landowner, or none of the other roles, clear patterns of

TABLE 3.1
Hiring Workers and Working for Others
(*from survey*)

Hires workers	Works for others	N	Percent of Farmers			
			Ages 25–34 ($N = 111$)	Ages 35–44 ($N = 100$)	Age 45+ ($N = 90$)	All ages ($N = 301$)
Yes	No	196	61%	67%	67%	65%
Yes	Yes	23	6	4	13	8
No	No	16	7	6	2	5
No	Yes	66	25	23	17	22

socioeconomic differentiation are reflected in the distribution of roles. Group leaders tend to be employers, and both group leaders and employers tend to avoid the worker role. Moreover, the distribution of farmers over the other roles available to them varies with age and size of operation (i.e., wealth).

Some 92 percent of the men from the hamlets of Nachig and Apas are farmers.[1] Table 3.1 shows the distribution of these 301 men over employer and worker roles, and includes a breakdown by age. Most farmers hire workers, and a good number work for others; but very few do both or neither. As might be expected, there is a slight tendency for the employers to come from the older farmers and the workers to come from the younger farmers, but the variation with age is not great. On the whole, these data do not support the idea that young farmers begin as workers and become employers when they are older. In fact, it would seem that Zinacanteco farmers divide themselves into employer and worker types early in their farming careers.

Data that I have not displayed in Table 3.1 show that group leaders are predominantly employers and tend to be older than the population as a whole. There are 27 group leaders who have groups with three or more members, and 25 of these are exclusively employers; the other two do not hire workers and work for others

[1] Of the 327 adult men from Nachig and Apas who were interviewed, 301 had fields in the lowlands. Details of this survey are given in Appendix A, pp. 163–68.

TABLE 3.2
Size of Operation
(*from survey*)

Hires workers	Works for others	Mean almuds seeded in the lowlands			
		Ages 25–34	Ages 35–44	Age 45+	All ages
Yes	No	3.1	4.2	4.6	3.9
Yes	Yes	—[a]	—[a]	—[a]	2.8
No	No	—[a]	—[a]	—[a]	2.1
No	Yes	1.6	1.8	1.9	1.7
Mean almuds seeded by age group		2.5	3.5	3.9	3.3

[a] The number of cases in these cells is too small to yield reliable means.

when their own work is done. The three ascending age categories shown in Table 3.1 include 37 percent, 33 percent, and 30 percent of the population respectively. But the 27 group leaders are distributed across the same categories as follows: 15 percent, 44 percent, and 41 percent. The youngest age group is drastically underrepresented, and there is essentially no difference between the two older groups. Although a farmer may employ workers early in his farming career, he is not apt to become a group leader until he has several years of farming experience.

Table 3.2 shows the amount of corn seeded by the farmers in each of the various employment and age categories. The most obvious pattern here is the relationship between hiring workers and seeding greater amounts. Zinacantecos generally regard two almuds as the most a farmer can handle alone, and even farmers with two almuds seeded usually hire workers during peak work periods.[2] Many farmers do not hire workers only because they have unmarried sons to help with their work. The breakdown of employers (the top row) by age shows a substantial difference between the amounts seeded by the youngest age category and the older ones. We may conclude, then, that a farmer does not expand

[2] The implications of farming operations of various sizes will be discussed at length in later chapters. The reader who wants an absolute measure may take an almud seeded to mean a bit more than a metric ton of grain harvested. Amount of corn seeded may be taken as a measure of wealth.

his operation fully until he has been farming for a number of years. And again, the younger employers are unlikely to have sons old enough to work, whereas older men often have the help of unmarried sons in addition to hired labor.

Group leaders are substantially larger operators than the population as a whole. Of the 27 group leaders mentioned above, the 25 who are employers average 5.2 almuds seeded, as compared with 3.9 almuds seeded for the class of employers as a whole. (The two who are workers seeded one and three almuds respectively.)

In sum, the typical Zinacanteco farmer is an employer who began hiring workers early in his farming career, and a few farmers are small operators who supplement their income from farming by working for others. By the time he has reached middle age a farmer may have become the leader of his own group; if he has, he is likely to run a very large farming operation.

Work Groups and Their Leaders

Most farmers who rent fields in the lowlands need never speak with a landowner. Instead, each farming group is represented by one of its members, the group leader (*baldio*). A single oral contract between the landowner and the leader covers all the land a group farms; and the leader concerns himself with subdividing the land among the members of his group and accumulating the rent from them at the end of the season. Thus most landowners need never speak with most of the farmers who work on their land. From the point of view of the landowner who may have dozens of Zinacantecos working on his land, and from the point of view of the farmer who may speak little Spanish and may be insecure about his own ability to protect his rights, this is an ideal arrangement. Since formal contracts are impossible, especially because landowners and farmers come from different cultural and geographic communities, the use of intermediaries reduces the number of people to whom any individual must relate, and it becomes possible to operate with personal oral agreements.

Membership in work groups is far from a random association,

and existing social relationships determine in large measure which farmers come together. Nevertheless, access to group membership is not considered a problem by Zinacantecos. Though a group leader may tell a prospective member who could in fact be accommodated that there is no more land available where he is farming, such incidents seem to be rare. And in some cases, leaders who have contracted for large expanses of land must actively seek out new group members.[3] Usually, a farmer who wants to join a new group seeks out the group leader, presents a gift of liquor or other delicacies, and asks for the amount of land he wants. One informant complained about a leader who demanded too many gifts; but in most cases these token presents, usually consumed on the spot, are neither a significant outlay for the farmer nor a source of economic advantage for the leader.

When the group actually reaches its land for the first time, the group leader has a more important economic function, for the land must be divided into plots that will give each group member the area he wants. Almost without exception, the land is hilly and of uneven quality. Normally, each farmer receives a strip including a mixture of land types; but in some cases equity is achieved by assigning each man two separate plots (see Stauder 1966 or Vogt 1969: 40–41 for further detail). Ideally, the group leader arranges for all members to be present when the measurements are made (see Appendix C, pp. 183–88).

Although some informants state that a group leader should be the equal of any member when the land is divided, it is clear that leaders tend to get better plots. One informant, a leader, explicitly stated that the leader has the right to take the best plot. Effectively, there is a tacit understanding that the leader will not suffer in the

---

[3] A group leader who tells an owner that he will take, say, fifty units of land is not responsible for all the rent if the members of his group eventually choose to farm something less than the total; but a leader who grossly overestimates the land his group will use endangers his relationship with the owner, who may be counting on income from all his land. This discussion may give the impression that the status of group leader is permanently held by some men to the exclusion of others, and that farmers are always attached to the same group. Actually, these alliances are constantly shifting, and many men enter and leave the status of leader every year.

division of the land; and on the few occasions when I got a clear picture of the division of actual fields, the group leaders had at least the advantage of being relatively close to the common camp-site of the work group. Some leaders grossly abuse their privileges: One informant told how he had once been assigned a smaller-than-average field, but had nevertheless raised a crop better than the group leader's. The next year the leader took that land for himself, assigning the informant an inferior plot. Gleefully, the informant reported that the leader's crop had been poor in spite of the "good" land. The story ends in the third year with the informant returning to his original plot and producing a good crop.

When the owner or his foreman is busy, the leader may be asked to divide the land and simply report the results to the owner. In this situation the leader is often able to get extra land for himself and the members of his group, although an obviously false report may bring the owner into the fields to make his own measurement. One group leader said that a leader should report fewer units of land than are actually measured and marked off and should then split the advantage in rent payment with the members of his group according to the units each member takes. Obviously, this procedure depends on both the honesty of the leader toward his members and the willingness of the owner to trust the leader's reports. Since the owner depends on the leader for good relations with the group, he may favor the leader even when he is supervising the measurement of the land. In one case a foreman "overlooked" the fact that a leader's strip of land was almost twice the standard length.

Thus, during the measurement and assignment of plots, a group leader often has the opportunity to gain advantages for himself relative to members of his group and for members of his group relative to widely recognized standards. But in general, Zinacanteco farmers do not seem to expect the advantages that may come at this stage of the work. They are more concerned with the protection that a leader can give his members by talking the landlord into a rent reduction when the crop yield is less than normal, and with other relations with the landowner that are negotiated by the leader. These protective aspects of the leader's role are discussed immediately below in the order in which they become relevant during the agricultural cycle.

After its land has been divided, a group must provide itself with a common shelter and must build or mend fences to protect the fields from roaming cattle. At this stage the group leader negotiates with the owner for building material; and if a new fence is needed, he settles the wages the owner will pay the group members who put it up. Ideally, this building and repair work is done in common, each farmer contributing according to the size of his plot.[4] Some informants complain about group leaders who administrated these arrangements unfairly. For example, leaders may avoid their own share of the work by saying they did it before others arrived; or they may assign fencing work in individual sections rather than in common and then blame a specific group member for damage to crops that occurs when cattle get through the fences at that member's section. The ideal group leader makes decisions and assignments concerning this work in a full meeting of the group.

A good leader will not commit the members of his group to working for the owner without their permission. Although the practice is less common than it once was, an owner may still try to recruit labor for his own fields by requiring that each renter put in some time there. Normally, renters are paid for this work, but the pay sometimes takes the form of a "gift"—for example, a cut from a steer to be slaughtered by the owner at the end of the season. One informant, in discussing the worst leader he had ever worked with, described a year in which he worked twelve days on the owner's field only to find that his share of the meat at the end of the season was about a quarter of kilogram (worth about a half day's pay). In this and other matters, the leader is supposed to protect his group from abuses by the owner.

In a normal year, each farmer delivers his share of the rent (in

---

[4] In addition, farming groups normally sponsor religious ceremonies to insure the success of their crops (see Vogt 1969: 455–61). These represent a small extra expense for group members.

corn) to the main ranch house after the crop is in. The leader arranges for this to be done on a single day or over a two-day period.
He then goes to the landowner and confirms that the total rent
paid by the group is satisfactory. It is the responsibility of the
leader to see that each man pays his fair share. Obviously, an un-
scrupulous leader could assign shares so that the total rent would
be provided without any contribution from him, but this seems to
be a rare occurrence. Although it seems to be expected that the
leader will gain some advantage in the selection of his land, he is
expected to pay rent on a par with others. At times, a group leader
will try to convince the owner to pick up the rent at the group's
campsite, thus saving them the trouble of delivery. As one informant put it: "A good leader will try to get the owner to pick
up the rent for the whole group even if the leader himself has
mules in hot country and can easily deliver his own share."

The group leader's role is critical when an individual or the entire group has a poor crop, since he is then responsible for nego-
tiating a lower rent. Zinacantecos recognize that owners differ in
their willingness to lower rent in a bad year; but good leaders are
skillful in convincing any owner that the rent should be lowered.
Since just what constitutes a normal crop or a poor crop is not
always clear, the leader may be very important to his group in
even a slightly bad year.

In sum, the group leader is the intermediary between the farmers and the landowner. The farmers in a group are economically
independent from each other and work as individuals, but the
landowner deals with them all through the leader. The ideal leader
is fair in dividing the land and administering the common work
the group must do to shelter itself and protect its fields. He makes
all arrangements in an open, above-board way. In relations with
the owner, the leader must protect group members from abuses,
and he must try to gain as many advantages as the owner will
tolerate. When yields are low he must renegotiate the original
rental agreement. In return for his services, the leader receives
small gifts and favors, the satisfactions of leadership, and the
knowledge that he will be able to choose the best among the many

plots that his group will farm. If he chooses to risk his reputation as a good leader and a fair man, he can often contrive to pay less than his fair share of the rent and do less than his share of the common work.

The Composition of Work Groups

The composition of work groups is subject to few limits. Size is determined by the inclinations of the men involved and the amount of land available to the group. Some men farm alone, and some groups have as many as twenty members. The norms for organization are almost universally those described in the section on group leaders. Occasionally, a landowner insists on a different arrangement; and men who deal independently with the landowner sometimes have adjoining plots and travel and live together, behaving like a work group except for their dealings with the landowner on allocation of land and payment of rent. On the whole, the groups that are formed are important alliances, though temporary and limited to farming, and they tend to reflect hamlet and kinship relationships that extend beyond work in the fields.

The great majority of Zinacantecos belong to groups lead by residents of their own hamlet. As Table 3.3 shows, in Apas only 20 of 136 group memberships are in groups with leaders from other hamlets.[5] The comparable figure in Nachig is 46 of 235 memberships. The principle that explains the overall trend also explains the difference between Apas and Nachig. Men tend to form groups with others who live near them; and whereas Apas is geographically isolated, the Nachig population extends at many points to meet the settlements of adjoining hamlets. This difference is also re-

---

[5] Group membership is counted from the point of view of the group, not the individual. The 39 Nachig farmers and 27 Apas farmers who belong to two groups are counted twice. In addition, the analysis of data on work-group composition includes three Nachig group leaders and one Apas group leader who are less than 25 years of age and are therefore excluded from most of the discussion. Thus the Nachig memberships include 193 farmers plus 39 double memberships plus three young group leaders, or a total of 235. The comparable figures for Apas are 108 plus 27 plus 1, or 136. Some of the groups shown in the table are, in actual practice, enlarged by members younger than 25 and by men from other hamlets.

TABLE 3.3
Work Group Size
(*from survey*)

Group size	Nachig		Apas	
	Number of groups	Total memberships	Number of groups	Total memberships
Groups with leaders from the same hamlet:				
1	15	15	6	6
2	5	10	12	24
3	10	30	—	—
4	2	8	3	12
5	5	25	6	30
6	1	6	3	18
7	6	42	—	—
8	1	8	—	—
9	—	—	1	9
12	1	12	—	—
15	1	15	—	—
17	—	—	1	17
18	1	18	—	—
TOTAL	48	189	32	116
Groups with leaders from other hamlets:[a]				
1	22	22	1	1
2	7	14	1	2
10	1	10	—	—
TOTAL	30	46	2	3
Special Apas group[b]	—	—	1	17
GRAND TOTAL	78	235	35	136

[a] The membership shown is probably a small fraction of the total group membership.
[b] This is the group working on land where the owner insists on dealing with each farmer separately. They are considered a group in other ways, but deal independently when paying the rent.

flected in kinship ties. In the population of adult males studied, the median number of relatives is six for Nachig men and ten for Apas men. There is no reason to suspect that Nachig men actually have fewer adult male relatives; rather, because of intermarriage with adjoining hamlets, many relatives of Nachig men are not included in my survey of Nachig. Nachig men who belong to groups with leaders from other hamlets tend to have fewer relatives in the survey population than do other Nachig men, and I interpret this to mean that they are farming with relatives in other hamlets.

Exactly how much kin ties determine the composition of work groups is difficult to state in a meaningful way. I found no satisfactory measure of the degree to which group membership varies from the extreme possibilities: a group in which every member is related to every other member, and a group in which there are no more kinship relations than would be expected from a random assignment of the population.[6] However, it would be startling if kinship were not important; and it is possible to show that it is. For instance, of the twelve Apas groups with two members, nine are formed by two relatives. Since the men involved have an average of 10.5 adult male relatives each and can choose their partner from 108 other Apas farmers, fewer than one in ten of the groups would include kinsmen if the association were not influenced by kinship in some way. In Nachig, the comparable figures show two of five two-man groups made up of relatives, and the mean number of male relatives for the men involved in the two-man groups is 6.5. Since the number of available associates in Nachig is 195, one would expect only one two-man group in about thirty to be made up of relatives. On the whole, kinship ties are important to group composition, but they are not the exclusive determinant of it.

Figure 3.1 shows the relationships of the men in a Navenchauc lineage defined by the senior male, Shun Vaskis (1 in the figure); and Table 3.4 shows the work groups to which the Vaskis men have belonged for the past ten years. Shun and his two sons by his second wife form the core group, which almost always works together under his leadership. One of Shun's four sons-in-law often joins them; but the other three (with the single exception of number 6 in 1960) head their own groups. Shun's son by his first wife remains independent from the core group, and is, as it happens, less involved in the lineage social life than any of the sons-in-law. The newest generation includes only one inmarrying male, Romin (9 in the figure), and he has joined the core group with his wife's

[6] In stating the extremes, I am assuming that kinship will not have an inhibiting effect on association in work groups.

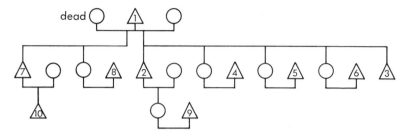

FIGURE 3.1. The Vaskis Lineage

father.[7] On the whole, the men of the Vaskis lineage so defined are wealthy and important members of the community who work independently; but the sons of Shun Vaskis by his second wife have stayed with him, and Romin has joined the core group, at least temporarily.

One of the Apas groups contrasts sharply with the small, kinship-dominated core group described above. The leader, Lorenzo, is an innovative and dynamic man in his thirties; he was raised by a widow who adopted him when he was orphaned, and he does not have strong ties with his brother, who was raised separately. The group in question includes Lorenzo and eight others. Although 58 percent of Apas men have at least one adult brother, only one of the eight besides Lorenzo has a brother. Lorenzo has no relatives in the group, and among the other eight men there are only two related dyads, a pair of cousins (brother's sons) and a pair of brothers-in-law. Lorenzo also heads a larger group that includes five brothers (who are unrelated to the rest of the group) and eleven other men (five of whom work with both of Lorenzo's groups). When the brothers are excluded, the 66 dyads formed by the other twelve, including Lorenzo, show only five kinship relationships.

In sum, work groups vary greatly in size. Men join and leave

[7] Romin's case is discussed on pp. 45–46 and in Appendix B. Names have been assigned to people mentioned in the text so that no two persons are referred to by the same name. Some names are fictitious. The Vaskis lineage, though resident in Navenchauc, is the one for which I have the most complete information (see also Cancian 1965a: 119–21).

TABLE 3.4

Vaskis Work Groups, 1957–66

					Lineage member				
Year	1	2	3	4	5	6	7	8	9
1966	A1	A1	A1	A1	D5	H6	N10	E8	A1
1965	A1	A1	A1	A1	D5	H6	N10	E8	A1
1964	A1	A1	A1	A1	D5	H6/I6	N10	E8	A1
1963	B1	B1	D5	B1	D5	J6	N10	E8	LX
1962	B1	B1	D5	B1	D5	J6	N10	E8	P9
1961	C1	C1	D5	C1	D5	J6	OX	E8	C1
1960	C1	C1	D5	C1	D5	C1	OX	E8	C1
1959	—	D5	D5	GX	D5	F6	OX	E8	D5
1958	—	E8	E8	GX	D5	F6	OX	E8	MX
1957	—	F6	F6	GX	D5	F6/KX	OX	E8	KX

NOTE: Each letter indicates a different named location in the lowlands. The numbers following the letters designate the group leader, as numbered in Figure 3.1, and an X in that position means that the group leader was not a member of the lineage shown in the figure. Individual 1 did not farm in the lowlands in 1957–59. Individual 6 worked in two locations in 1957 and in 1964.

them as individuals, though a core group may remain together over the years. Most men join the groups of leaders from their own hamlets, and kinship is often influential in determining how a group will be formed.

Landowners

The owners of the lowland fields on which Zinacantecos farm are a diverse group. Land reform broke up the great estates that existed in the area at the beginning of the century and left a multiplicity of smaller ranches. The larger holdings that remain are run as cattle ranches, for the reform laws allowed more land to a rancher than to a farmer whose land was arable. The smallest holdings belong to members of ejido communities, all of whom received their land as a grant under the reform laws.[8] The ranches that rent fields to Zinacantecos range from large, modern, mechanized operations to holdings that barely allow the owner to have a two-room house and avoid working his own land. Many owners,

---

[8] Although ejido members are not legally permitted to rent their land, Zinacantecos occasionally rent from them for short periods.

in fact, work in the cities and visit their lands only occasionally; in these cases the ranch manager, often a poor relative who farms himself to supplement his income, is the main contact of the Zinacanteco farmer. I have no data on the size of different ranches in hectares, but a description of representative administrative arrangements may give the reader a better idea of what is involved, especially since the land is extremely variable in quality.

The largest ranch I have seen in the area is owned by a man who has many other commercial interests and is regarded as one of the richest men in the state. The main house, which was no doubt built as a vacation retreat as well as a place for him to stay on his regular visits, includes four bedrooms, bathrooms, and a dining room that will seat twenty without crowding. Guest houses are spread around the property, and the ranch has its own water and electric systems. The administrative staff includes a professional manager, a bookkeeper, and the owner's son. A fourth man is in charge of the veterinary medicines, the warehouse, and a short-wave radio that is used to communicate with the owner's other business locations. When he is not there the owner radios his son or his manager twice a day; the radio is also used to order supplies and ask medical advice. Ranch equipment includes some half-dozen cars and trucks and dozens of other machines. The ranch's regular employees include drivers, cowboys, and foremen for temporary work crews, which are principally recruited from nearby ejido communities whose members work when their fields do not demand attention.[9]

I first visited this ranch in 1967, with a group of about ten Zinacantecos who planned to measure out their land and prepare it for seeding on a later trip. They had not farmed there the previous year and were not sure that they would be given land. They went first to one of the ejido communities near the ranch to visit the work-crew foreman who was likely to supervise the measurement

[9] During one of my visits to this ranch (spring 1967), hired earth-moving equipment was being used to drain new pasture land. The owner planned to expand his dairy herd to meet the anticipated demands of a powdered milk and chocolate plant that was to be built in Chiapa de Corzo. If the plant is ever completed, milk production may become very profitable; and in the long run the land available to Zinacantecos will probably diminish significantly.

5. A lowland shelter during the harvest season.

6. Zinacantecos trying to recruit a Chamula worker on the outskirts of San Cristóbal.

7. Small landowner's house on the new road from Chiapa de Corzo to Acala.

8. Large landowner's house on a ranch in Zone 5.

of the land. After presenting gifts of liquor, fruit, and cigarettes, they asked that the foreman recommend them to the ranch manager as more rightfully entitled to the land than another group they knew to be competing for it. The foreman's family welcomed the group and allowed them to cook some meat in the house—a three-room wattle and daub affair with a floor of packed dirt.

We slept in the foreman's yard and left in time to reach the ranch for the morning shape-up of workers. As it turned out, the manager granted the land to the group, gave them a measuring rope, and sent them off to measure their own plots; the foreman was sent off with his crew to do other work. I later learned that the owner planned to use the land for pasture the following year and had given the Zinacantecos a discount on rent in exchange for their seeding a forage grass between the rows of corn. If the grass took hold the land would be taken out of corn production in the following year. (The land, which had lain fallow for two years, is shown in Photo 10, p. 57.)

Other landowners operate on an entirely different scale. A nearby ranch is managed by a young man and his widowed mother, who are the custodians of family land in which they and a number of others hold shares; the son had been a minor bureaucrat in a San Cristóbal office until about a year before my visit to the ranch. The two live in a large, crumbling one-room house surrounded by a veranda. A fenced-off section of the yard is filled with scurrying chickens and a couple of pigs; water is brought to the house by pack animals; and the bathroom is a roll of toilet tissue and a walk to any discreet location away from the house. The land and house, a few cattle, and a recently acquired secondhand pickup truck that may not pay for itself represent the total capital of this ranch. A few resident families exchange work in the ranch's fields and other services for plots on which they can grow their own crops. The son hires a tractor to plow the limited area he farms himself, and is now wondering how he can get the capital to develop milk production (in anticipation of the new milk processing plant to be built in Chiapa de Corzo).

A Zinacanteco farmer is not particularly concerned with the size of a ranch, for any of the lowland ranches is large enough to accommodate several work groups of average size. Taking each separate farming location of each Nachig farmer for each year during 1957–66 as an individual case, my survey gives a total of 1285 "farmer-year-locations."[10] About 70 ranches were used by at least one Nachig farmer for at least one year, but four ranches account for 49 percent of the total. The fourth largest, which accounted for about 10 percent of the total (127 of 1285 farmer-years), was the smaller of the two ranches just described. From the Zinacanteco point of view, then, almost any ranch is big enough. Another of the lowland ranches had just a handful of Nachig farmers in 1957–66 but accounted for 378 Apas farmer-years (53 percent of the Apas total) during the same period. For reasons discussed in the next section, this probably represents the most extreme concentration of farmers from one hamlet on one ranch. Each hamlet tends to concentrate on a few ranches—as might be expected, since the work groups are usually formed on a hamlet basis and news of available land usually travels the same way.

In sum, most farmers have little contact with landowners and care little about landowners' affairs except as these affect the farmers. Though ranches vary greatly in size, virtually all are large enough to accommodate more than one large group of Zinacantecos. From the farmer's point of view, then, almost any ranch is big enough and any landowner who is not malicious is good enough. Group leaders care more about the characteristics of landowners, for they have to deal with them.

Rent

Rent is typically paid in kind, at the rate of 24 almuds per almud seeded.[11] There are a variety of exceptions and special cases, but in

---

[10] In my calculation of farmer-year-locations, farmers who work in two locations are counted twice. There are actually 1190 farmer-years for Nachig, including 95 farmer-years with two locations, making a total of 1285.

[11] My arbitrary use of the almud as the single most convenient unit of measure gives a slightly unreal cast to this description. Rent is "really" two *fanegas* per almud seeded (1 fanega = 12 almuds), and no Zinacanteco would describe it in any other way.

TABLE 3.5
Yields and Rents by Zone: Rent Paid in Full
(*from work histories*)

Zone	Farmer-years	Contracted rent[a]	Mean rent	Mean yield	Lowest yields		
					1	2	3
1	12	8–18	13	80	28	46	60
2 (low)	15	13–18	16	82	60	60	68
2 (normal)	41	24	24	96	48	64	66
3–7[b]	25	24	24	137	78[c]	96	96
9	2	32	32	198	180	216	—
4	5	36	36	145	78	138	150

[a] All rents and yields are in almuds.
[b] The two cases in Zone 7 have yields of 115 and 156.
[c] This is the one case from Zone 3.

the end a farmer usually pays 20–25 percent of his harvest as rent. I will first discuss cases in which the rent agreed on before seeding was paid in full, then cases in which the crop was exceptionally poor and the landowner accepted less than the originally agreed upon rent, and finally special arrangements that affected only a few farmers. I will deal with 144 actual cases reported during the work history interviews.[12]

Table 3.5 displays the 100 cases in which the full contracted rent was paid. The majority of these show the typical 24-almud rent. Those that involve a lower contracted rent are all in Zone 1 and the parts of Zone 2 that border on Zone 1 (see maps). In some of these cases a low rent was actually agreed on beforehand. But quite often the lower rent shown in the table results from an arrangement under which the landowner is "generous" in measuring the area of land to be seeded. When 1⅓ almuds of seed enter such

12 Eighteen of the twenty informants whose work histories were taken gave relevant information. Their reports for 1957–66 cover 180 potential farmer-years, during 32 of which no corn was seeded in the lowlands (these are mostly years before the younger informants became independent farmers). Twelve farmer-years were added to the 148 remaining by informants who farmed two fields in one year. These 160 farmer-years break down as follows: 100 in which rent was paid as contracted (Table 3.5); 18 in which rent was lowered (Table 3.6); 26 in which unclassifiable arrangements were made; and 16 with incomplete information. Collection of work histories is discussed in Appendix A, pp. 163–68.

a "generous" one-almud plot, rent that is set at 24 almuds effectively becomes 18 almuds. The farmer, who is concerned with the amount he can seed, and the landowner, who measures the area of land, bypass the usual equivalence of two measurement systems and produce a compromise that makes the relatively infertile land of the area more attractive to the farmer. In Zone 9, where the land is excellent and the actual rent is 32 almuds, a similar process works in the opposite direction. The rent is usually stated as "two fanegas," as it is in the other zones; but the measuring system customarily used for rent payments in Zone 9 has 16 rather than the usual 12 almuds in each fanega, and both parties understand this. The five cases of 36-almud rent in Zone 4 come from a single ranch, where that rate is charged for level fields that Zinacantecos seldom rent and the normal 24 almuds is charged for the normal hilly fields. On the whole, rent varies with quality of the land as indicated by the mean yield, and on the average it is less than 25 percent of the crop.

Table 3.6 is a better illustration of how the expectations of farmers and landowners about rent and yields vary from zone to zone; it displays the eighteen cases in which rent was lowered because the crop was extraordinarily poor. In Zones 4 to 6, where there were no contracted rents of less than 24, rent was lowered for a yield as high as 90 almuds; but in Zone 1, where the mean yield does not even reach 90 almuds, the highest yield considered worthy of a lowered rent was 42 almuds. The corresponding figures in

TABLE 3.6

Yields and Rents by Zone: Rent Lowered for Poor Crop
(*from work histories*)

Zone	Farmer-years	Mean contracted rent[a]	Mean paid rent	Mean yield	Highest yields		
					1	2	3
1	6	14	8	37	42	36	36
2	5	18	12	53	76	60	48
4–6	7	24	14	64	90	80	72

[a] All rents and yields are in almuds. One case in Zone 1 and one case in Zone 2 had a contracted rent of 24; all contracted rents in Zones 4–6 were 24.

Table 3.5 show full contracted rent paid for a crop of 28 almuds in Zone 1 and 96 almuds in Zones 4–6.[13] Also notable in the tables is the fact that cases of serious crop failure were far more common in Zone 1 than in other zones.

The 26 cases that do not fall into the categories used in Tables 3.5 and 3.6 illustrate the diversity of situations farmers face. A simple list of the thirteen special situations they represent follows.[14]

Three cases involve payment of at least part of the rent in cash. One includes cash in advance, labor for the landowner during the year and corn delivered at harvest time. The second involves simple cash payment in advance. And the third comes from a ranch in Zone 8 where the owner takes cash at harvest time—apparently in order to avoid the extraordinary cost of getting corn from his ranch to the nearest road. In two cases owners refused requests to lower rent (for a yield of 60 almuds in Zone 2 and a yield of 78 almuds in Zone 6). In two other cases group leaders paid extraordinarily low rents on good yields because of special arrangements made with the owners.

One owner in Zone 1 charged no rent because he wanted the land cleared for pasture and planned to graze his cattle on the cornstalks after the harvest (yields for two years on this land were 72 and 56 almuds). An owner in Zone 6 charged no rent after his cattle broke into the fields and reduced the harvestable corn to 48 almuds per almud seeded; and the same man twice accepted full rent in rotted corn when the fields were flooded by a nearby

---

[13] The Zone 2 figures show relatively low minimum yields in Table 3.5 and relatively high maximum yields in Table 3.6. This anomaly is explained by the difference between one important ranch in Zone 2 (the Flower Ranch) and all the other Zone 2 ranches included in the tables. All three of the lowest yields for Zone 2 (normal) are from this ranch; and my complete data show no rent lower than 24 almuds for the Flower Ranch, even though it is represented by 23 cases. Zinacantecos are fully conscious of this ranch's "tough" policy on rent, but continue to farm there because of other advantages it offers.

[14] In addition, one informant who did not produce a work history reported paying 60 almuds rent for a piece of land that took 1⅓ almuds of seed (i.e., a rate of 45 almuds per almud seeded). This was flat, rich land that had been pasture for a number of years, and the Zinacantecos working it hired local people to plow it with ox teams. They got an extraordinary yield of 162 almuds per almud seeded, plus a full bean crop see Photo 23, p. 186). Thus the arrangement was profitable in spite of the high rent.

river.[15] The final three cases involve an owner who prefers to set rent only after the harvest is in (he charged 12 almuds for a 60-almud yield), an owner who misjudged the extent of a farmer's land and charged too little, and a much disliked group leader who fined a member of his group an extra six almuds for appearing late to pay the rent.

Rent is the most important part of the economic exchange between the farmer and the landowner, and the crop yield and work input on a piece of land are the most important factors in a farmer's judgment of whether he is getting a good bargain. However, unusual renting arrangements sometimes tip the balance between attractive and unattractive locations. This is true of the Apas farmers who work at the Flower Ranch. Although the land produces relatively low crop yields, the ranch offers three substantial advantages.

First, the Flower Ranch is only six hours walk from Apas, and few other locations are so close. Second, the wife of the manager speaks Tzotzil, so that even monolingual group leaders can communicate effectively with an official representative of the landowner. Many lowlanders, including landowners, speak a few words of Tzotzil; but dependable communication in that language is impossible on most ranches, and farmers usually have to rely on a group leader who speaks Spanish. Third, and perhaps most important for many Apas farmers, the Flower Ranch reserves certain fields for grazing the mules of farmers who rent land there, and this pasture is superior to any near the hamlet of Apas. One nearby ranch charges ten pesos per month per mule for similar grazing because its pastures are normally used for the owner's cattle, and farmers are not even allowed free grazing when they are there to work their fields.

In exchange for their grazing rights on the Flower Ranch, Apas farmers used to make four trips to San Cristóbal each year to deliver the owner's own corn and fruit crops to market. Since the new road has opened, corn has been sent by truck, and the farmers

---

[15] These cases involve the farm run by a young man and his mother that is described in the section on landowners.

have had to take only two mule loads to San Cristóbal for each mule that grazed during the year. They are paid roughly half the going rate for this service. In effect, they are paying one peso per mule per month of grazing. Before the road opened, this exchange provided the owner with secure, inexpensive transportation and the farmer with superior pasturage at no out-of-pocket cost. Apas farmers still use mules extensively, since Apas is close to the lowlands and cannot be reached by car. However, the importance of pack transport is decreasing in the region as a whole, and for this reason the farmers' opportunities to earn cash by contract hauling when there is no work in cornfields are also decreasing. Thus this special feature of the Flower Ranch renting arrangement, like many other special arrangements, gives farmers advantages that tie them to a system that is disappearing as the region modernizes.

In sum, most farmers agree to a rent in kind at the beginning of the season and pay it at harvest time. Rent is typically 24 almuds per almud seeded, but it tends to vary with the quality of the land and usually amounts to 20–25 percent of the crop. When yields are poor the owner normally lowers the rent, thus sharing the loss with the farmer. Though rent, necessary input, and yield are the principal factors determining the attractiveness of a given plot, some ranches offer special arrangements that tip the balance for some farmers. For the calculations to be made later, I will set standard rents at 12 almuds in Zone 1, 24 almuds in Zones 2–8, and 32 almuds in Zone 9. I will further assume that rent is lowered to 25 percent of the crop if yields are not more than four times the contracted rent.

Changing Locations

Most lowland fields are exhausted by repeated use, and farmers who want optimal yields must move to a new plot every few years. Zinacanteco farmers know this, but their actual decisions on moves are influenced by a complex of factors, including yield in previous years, the labor input needed to maintain yield, the opportunities for renting better land elsewhere, and the landowner's willingness to have them continue working on the land they have.

TABLE 3.7
Moves per Farmer, 1957–66
(from survey: primary plots only)

Hamlet	No. of farmers	Number of moves				
		0	1	2	3	4–7
Nachig	119	24%	46%	18%	10%	1%
Apas	62	37	27	18	13	5

Table 3.7 shows the number of moves made by Zinacantecos who worked in the lowlands for the entire period 1957–66. It includes one location per farmer for each year (secondary locations will be discussed below). According to these data, Nachig farmers spend a mean of 4.6 years on a ranch before moving; the corresponding figure for Apas farmers is 4.4 years. As the table shows, Apas farmers are more widely spread around their mean than are Nachig farmers. A substantial proportion of Apas farmers did not change ranches at all during the ten-year period, and 87 percent of these worked the entire period at the Flower Ranch. These data provide the most accurate overall picture of change in locations. However, since they include only movement between ranches, and since many farmers move to new plots on the same ranch every few years, the table may underestimate changes motivated by the prospect of better yields. On the other hand, the plot that is "new" for one farmer may have been worked by another man a few years previously or even in the year just past. Given the data available, it is impossible to treat movement from location to location in terms of soil fertility alone.

Some farmers farm two locations in the same year (Marian, described in Appendix B, for example). Nachig farmers had such plots in 8 percent of the 1190 farmer-years considered in Table 3.7, and Apas farmers in 14 percent of their 620 farmer-years. Since the table lists only primary locations (defined as those on which a farmer spends the most time), it is conservative as an estimate of all moves actually tried by Nachig and Apas farmers. If the secondary locations are added, the mean number of years on a ranch before moving drops to 4.4 for Nachig and 4.1 for Apas. The table also disguises a few isolated cases of very frequent moves.[16] The most dramatic is that of an Apas man who maintained a primary plot at Flower Ranch and tried eight different locations during the eight years that he had a secondary plot.

The story of the moves made by Romin, of the Vaskis lineage, will illustrate the variety of factors that influence farmers. In 1956, when his father died, Romin had been farming with him at a Zone 3 ranch on the far side of the Grijalva River. He left after harvesting and selling the crop because he did not want to work so far from home. His father had always sold their crop at the ranch, since transportation to their home was impractical; but Romin did not want to continue the arrangement. For 1957 he moved to a ranch just across the river. The group leader was from another hamlet, but Romin heard of the land from an uncle of his future wife (6 in Figure 3.1). The entire group left these fields at the end of 1957 because the owner refused to lower his 24-almud rent when they had a bad crop (45 almuds per almud seeded).

The next year Romin got land with a group leader from his own hamlet; he heard about this land from his maternal uncle, who worked in the same group. These fields, too, were abandoned after one year, this time because the owner turned the land into pasture. In 1959 Romin worked with another of his future father-in-law's relatives (5 in Figure 3.1), having heard from his future father-in-law that land was available. He left there to join the group of Shun Vaskis (1 in Figure 3.1), who had found land at another location. In this year (1960) he became formally engaged to Shun's grand-daughter. The group left this location after the second year because the owner had decided to collect the rent using a new almud measure, which the farmers calculated held about 16 percent more than it should have.

The group dispersed and Romin got a piece of land alone. He left at the end of the year because his yield was only 27 almuds per almud seeded and he had to pay one-third of this in rent. The next year he worked with another group leader from his own ham-

---

[16] The table would also disguise moves away from and then back to a single ranch, but this seldom happens.

let. But his yield was only 48 almuds per almud seeded, so he left at the end of the year. The group leader and other members of the group had better pieces of land, so they stayed. By this time (1964) Romin's future wife's grandfather, Shun Vaskis, had found the land he is still farming (see Photo 23, p. 186), and Romin rejoined the core group of the Vaskis lineage.

Other informants gave similar reasons for changing ranches. Some described the condition of the land rather than the yield in giving reasons for leaving: that is, the land was too stony for easy weeding; or it was too hilly, so that the topsoil was washing away. Others had trouble with dishonest group leaders or with owners who would not lower the rent in a bad year. An owner's decision to farm land himself or put land into pasture was also a common reason for changing location. All in all, farmers decide to change location for a variety of reasons. Quite often a strong component in the decision to move is the feeling that the crop yield expected for the coming year is less than desirable given the other aspects of a farmer's situation.

So far, I have emphasized the Zinacantecos' reasons for changing location; but there are many farmers who have similar experiences and do not move. For them, we must assume some kind of calculation that makes moving undesirable even when they experience difficulties (as in the case of the many Flower Ranch farmers whose rent is not lowered when their crops are poor). It is, of course, harder for informants to make articulate statements about why they do not move than why they have moved. My impression is that reasons for not moving can be divided into two types. First, moving to a potentially better location may involve noneconomic costs that the farmer is unwilling to pay. For instance, he may be part of a large kinship group that is unwilling to move. Or, like one informant who stayed in a relatively poor location so that he could be close to Zinacantan when he was called on to take part in religious ceremonies, the farmer may have noneconomic activities that are more important to him than achieving an optimal arrangement in his economic activities. Second, the uncertainty of any new, unfamiliar location may deter many farmers from moving.

In sum, Zinacanteco farmers move to new ranches for a variety of reasons, including difficulties with landowners or group leaders and the movement of kinsmen or others with whom they customarily farm; but the prospect of unsatisfactory yields is perhaps the greatest single factor behind movement. Noneconomic factors, however, are often crucial in decisions to move or stay. On the average, farmers move from one ranch to another every four or five years.

Workers and Employers

Zinacanteco farmers do not make elaborate distinctions between types of workers, although they look for the most reliable, hardworking men they can find. In recent years, labor has been in short supply at the peak work periods, and most farmers are delighted if they can recruit all the help they need. (Details on the recruitment of workers are given in Chapter 4 and Appendix B.) Older farmers sometimes say that years ago men worked harder and the workday was longer. Given the increasing prosperity of the region as a whole, there is good reason to believe that this is more than simple selective memory on the part of an employer who depends on the work of others for his profits. But since this trend does not distinguish between good and bad workers today, it is of little importance here. At present, work crews normally begin shortly after sunrise and take a substantial break in late morning for rest and a meal. They finish for the day well before dark, collect wood and water, and cook their evening meal. I have too little information to make a reliable estimate of the typical workday; but for the comparativist who must have such an estimate, eight hours is close to the average.

Since they are primarily employers, Zinacantecos are much more concerned about the employer role than the worker role. Most farmers who hire workers enjoy discussing the employer role, and they stress the need to treat workers well if one wants to have a reliable supply of them. Certain features of the "contract" between worker and employer are almost invariable:

1. When the worker is recruited the arrangements are confirmed by a gift to the prospective worker. In the overwhelming majority of cases this is liquor, which the employer and worker consume

together; but occasionally it may be some other luxury, or even an appropriate amount of cash explicitly given in lieu of liquor.

2. The worker is fed by the employer from the time they leave the highlands until the time they return to the highlands.

3. On return from the lowlands the employer provides a full meal (including meat and liquor), usually at his home.

4. The worker is paid for days spent working, but not for days spent traveling to and from the lowlands.

5. If the worker is paid in corn rather than cash, the employer delivers the corn to the worker's home.

6. If the employer and worker travel by bus or truck, the employer pays all costs.

The ideal employer is generous in his provision of gifts at recruitment, food in the lowlands, and food at his home after returning from the lowlands. The staple foods on the job are dried tortillas, ground cornmeal that is stirred into water and drunk, beans, and·greens that may be gathered by the workers. Besides salt and chile peppers, which he routinely supplies, the employer may also buy luxuries like tomatoes, potatoes, melons, salted shrimp, and meat; an occasional cigarette is also considered desirable. After returning from the lowlands, workers who live far from the employer's house often sleep there while recovering from the trip and the feast provided by the employer. The details on Antun's recruitment and maintenance of workers (Appendix B) illustrate the standard pattern, although Antun's expenses for extra food are relatively high.

Most of the exceptions to these generalizations occur in the employment of Zinacanteco workers with particularistic ties to the employer, and these cases tend to occur at points in the agricultural cycle when intense work is not required and many workers are available. In one case, a farmer found that he did not have enough corn to feed the group of workers he planned to take to the harvest, and he decided to get what he needed by harvesting some of his crop early. Since he did not have time to go to the lowlands himself, he asked a friend to do it for him. (This friend was also a neighbor, a *compadre*, and his stepmother's brother's son.) The fields were near a road, and the whole job took two days, including travel and work time. The farmer paid all expenses, plus two full days of wages. Another farmer, the leader of a large work group, found that he was occupied with the duties of a religious office at the time his fields needed "doubling" (see Chapter 4). Two members of his group were going to the lowlands to double their own relatively small fields. (One was the compadre of the group leader, since the group leader was godfather to all three of his children.) The group leader asked these men to double his fields for him and paid one-way transportation and regular wages for the days they worked.

Sometimes the farmers in a group finish their work at different times. In this situation, one farmer may work a day or two for another so that the second can finish quickly and they can travel to the highlands together.

In sum, most Zinacantecos are employers and are concerned about the employer role rather than the worker role. The normal "contract" between employer and worker permits the worker to reach his home with his formal wages intact, for the employer provides maintenance and pays for transportation. Typical relations with workers are universalistic; but smaller farmers may recruit workers on the basis of particularistic ties, especially for jobs done when labor is plentiful.

The Cost of Workers

Workers may be paid in cash or in corn. They tend to prefer corn at the middle of the agricultural cycle, when corn is scarce and expensive, and cash at the end of the cycle, when corn prices are low; but there is no substantial shift from season to season. In 1966 the mean wage in corn was 5.2 almuds per six-day week, and the mean cash wage was $45.20 per week.[17]

The two types of payment are very close to equivalent, depending on how one makes the conversion from corn to cash. The stan-

[17] In 1957 the mean corn wage was about 4.5 almuds and the mean cash wage about $30.00. The increase in cash wages represents inflation in part, but the increase in corn wages is real. These changes are discussed at length in Chapter 7. Since I am trying to estimate relative profits from the various alternatives open to farmers in 1966, I will take the 1966 wages as standard. The mean wages for 1966 are based on reports from work histories; eleven for corn wages, and ten for cash wages.

dard conversion figure used in this study is $9.00 per almud, making the corn wage worth $46.80. At the mean San Cristóbal market price for 1966 ($8.85 per almud) 5.2 almuds were worth $46.00. The cost of delivering the corn to the worker's home raises the corn wage substantially above the cash wage. On the other hand, if we make the reasonable assumption that cash is gotten by sale of corn, then the expenses of sale should be added to the employer's costs.[18] Altogether, a mean wage for 1966 is difficult to fix. Given the uncertain costs of delivery to the worker's home and sale by the worker, the best estimate is probably $50 per week, or 5.2 almuds of corn plus $4 per week.

Although I have set this standard wage for all workers, pay does vary. Zinacantecos sometimes explain higher than normal wages by the fact that the ranch in question is relatively far from the highlands. They say workers are reluctant to go so far and hence demand more pay. We might also expect farmers who run large operations to pay higher wages, for they must attract many workers and cannot afford to depend on particularistic ties to recruit men who will work for low wages.

Data from the work histories were used to test these hypotheses.[19] As can be seen in Table 3.8, corn wages and distance are not associated; and, if anything, higher cash wages are paid in nearer rather than in more distant locations. There seems to be a tendency for larger operators to pay the modal five-almud wage rather than a higher or lower one when they are paying in corn,

[18] In Appendix B the tables show that Marian spent two days with his horse (short days, he reported) delivering 30 almuds to his Chamula workers, and that he also paid 30 almuds to Hteklum men with no significant transportation cost. Romin had delivery costs of $0.67 per almud for 35 almuds paid to Chamula workers, and no expense for 16 almuds paid to Zinacantecos, making his mean cost $0.47 per almud. Antun, who used Chamula labor only when he paid in corn, had a delivery cost of $0.74 per almud. The expenses of selling corn are discussed in Chapter 5.

[19] In the 148 farmer-years reported in work histories (see Note 12), no workers were hired during 29, lowland people were hired as workers during two, and fields were farmed on a cooperative basis with relatives during eleven. In the remaining 106 farmer-years Indian workers were hired; to these were added six farmer-years in which a farmer hired workers for two locations. The slight deviation of sample size in Table 3.8 from these figures, appropriately modified by information given in the text, represents cases that are hard to classify.

TABLE 3.8
Wages, Work Location, and Size of Operation
(*from work histories: all figures in farmer-years*)

Wage	Location by zone[a]			Almuds seeded[b]		
	2	4–5	6–9	≤2.7	3–3.4	≥3.5
Corn wage in almuds/week:						
>5	12	6	5	16	3	3
5	31	13	7	18	15	15
<5	13	9	7	17	6	6
Cash wage:[c]						
High	22	7	6	8	12	12
Low	11	7	9	12	8	6

[a] 112 farmer-years considered; $N = 103$ for corn wage and 62 for cash wage.
[b] 106 farmer-years considered; $N = 99$ for corn wage and 58 for cash wage.
[c] 1957–63 high ≥$36, low ≤$30; 1964–66 high ≥$42, low ≤$40.

and to pay relatively high wages more often when they are paying in cash. Given the dispersion of what trends there are, the small number of informants involved, and the problems of accumulating their reports over the years, it seems reasonable to conclude that there is no real difference of the kind we are looking for; hence it is not worth trying to set different wage scales according to location of fields or size of operation.

The data presented in Table 3.8 suggest two other conclusions. In more than 90 percent of the farmer-years considered, corn wages were paid; cash wages were paid in about 55 percent. Insofar as this sample is representative, we may say that virtually all farmers who hire workers pay in corn part of the time, and that almost half of them pay exclusively in corn. The table also suggests that larger operators pay cash wages more often than smaller operators. The mean amount seeded where cash wages were paid was 3.4 almuds; where only corn wages were paid it was 2.8 almuds.

The basic cost of feeding workers in the lowlands is essentially the same for every employer; but estimating the other costs that must be added to wages is a complicated matter (see Appendix B). The problems may be reduced to three: (1) Although the size of the recruitment gift and the meal on return may vary somewhat with the length of time spent working, the variation is smaller than the variation in length of work trips. Thus a farmer who takes short

TABLE 3.9
Maintaining Workers
(*pesos per man-day worked*)

Expenditure	Marian	Romin	Antun
Corn and beans	$1.30	$1.30	$1.60
Extra food	—	.15	1.30
Recruitment	.20	.10	.15
Meal on return	1.10	.75	.80
TOTAL	$2.60	$2.30	$3.85
Man-days worked per worker per trip	4.5	8.7	9.2

NOTE: This table is based on the tables in Appendix B. For corn and beans calculations, both farmer's and workers' days are counted, with travel days weighted one-half because unreported food is eaten at home on travel days. For the other items, only workers' actual workdays are counted. Antun runs the largest single operation.

trips is at a comparative disadvantage. (2) The farmer who runs a large operation and depends on recruiting large numbers of anonymous workers must give more extra rewards than the small operator who can depend on kinship and personalistic ties for most of his workers. (3) The large farmer may also find it necessary to hire someone to help his wife prepare the workers' food for consumption in the lowlands, whereas the small operator can usually depend on his wife's "free" labor.[20]

Table 3.9 summarizes some figures that illustrate the first two problems. Since I do not have statistical data that would provide reliable means, I will set figures intuitively within the constraints suggested by the data in Table 3.9, by many bits of similar data, and by my general knowledge of the Zinacanteco farming situation. The variance in cost of recruiting and maintaining workers seems great enough to justify my breakdown by type of farmer and situation; but its ultimate economic importance is too small, I think, to justify a detailed discussion of all the factors that go

[20] Obviously there is much variation that cannot be predicted by scale of operation. A lazy woman in a household without other adult women is less apt to produce food than an industrious woman with many grown daughters to help her. Women who have no adult male in the household, commonly widows, often seek work preparing the dried tortillas taken to the lowlands by farmers. They are usually paid in corn, at three or four units for each four units (separately provided by the farmer) that they make into food. A major part of this job is gathering the wood used in cooking the tortillas.

TABLE 3.10
Estimated Costs for Recruiting and Maintaining Workers
(*pesos per man-day worked, to nearest $0.05*)

| | Number of workdays per trip | |
Number of workers	6	12
1	$3.15	$2.60
2	3.15	3.10[b]
3	3.65[a]	3.75[c]
4	3.90[c]	3.85[c]
5	4.05[c]	3.90[c]
6	4.15[c]	3.90[c]

NOTE: This table is based on the estimates shown below.
[a] Includes big-operator cost.
[b] Includes pay for food being made.
[c] Includes big-operator cost and pay for food being made.

into setting the figures. Thus, they may be simply stated:

Basic maintenance, including corn, beans, chile, salt, and extra foods: $1.50/day[21] + $3.00/trip.

Recruitment gifts: $0.10/day + $1.00/trip.

Meal on return: $0.40/day + $3.00/trip.

Extras for big operator: $0.50/day when three or more workers are used at one time.

Pay for food being made: $1.00/man-day worked in excess of 24 per trip (including both workers and farmer).

In sum, the cost of workers involves wages, recruitment costs, and maintenance (transport costs are discussed in Chapter 5). Wages may be calculated at $50 per week; basic recruitment and maintenance costs are $2 per worker-day plus $7 per worker-trip. The farmer who must hire more than two workers is estimated to have an additional cost of $0.50 per worker-day, which the smaller operator can avoid by recruiting through particularistic ties; and the farmer whose fields require more than 24 man-days during any single trip is assumed to spend $1 for food preparation for each extra man-day more than twenty-four. Table 3.10 gives some examples of total costs per man-day for different numbers of workers and trips of different lengths. All figures assume that the farmer himself works with his employees.

[21] Travel time is not included in the calculation of worker-days, but maintenance during travel is covered by the $3 constant cost.

FOUR

Labor Inputs and Yields

THE PRODUCTION process involves five essential steps: preparing the land, seeding, weeding, "doubling" the mature plant, and harvesting. The time required for seeding and doubling varies very little from place to place and farmer to farmer; but the time required for preparing the land and weeding varies with the type of land. The time required for harvest varies with the size and quality of the crop. In this chapter I will give a brief technical description of each step and will discuss the factors contributing to variation in labor input at each step; I will also describe input not directly involved in production, the use of workers by farmers who seed more than they can work themselves, and the yields in the various zones.

As is generally true when technology is simple and investment in labor is great, there are no significant economies of scale in the Zinacanteco production system. Thus I have expressed labor inputs in terms of a single basic unit, and inputs for larger units may be derived by simple multiplication. The basic unit may be expressed as either volume of seed or area of land; for, although there is some variation, a tablon of land (about two-thirds of a hectare) is normally planted with an almud of seed (15 liters volume). This land/seed equivalence is basic to Zinacanteco thinking about corn farming. The variance of both measures and the equivalence that occurs in practice are discussed in Appendix C.

Slash and Burn Agriculture in the Grijalva River Valley

In slash and burn agriculture the farmer ideally begins with an area of mature secondary growth, cuts down the trees and underbrush, waits for them to dry, burns the fields over, and finally uses a dibble stick to make holes for his seed. He may take only one crop from the land and then move on to another place and another stand of mature forest, but usually the effort of clearing land is not repaid unless he stays longer than one year. Often, he will stay on a newly cleared field for four or five years before moving. Conklin (1961: 27) has defined slash and burn or swidden agriculture (more generally, shifting cultivation), as any system in which the fallow periods are longer in years than the cropping periods. In many parts of the world, mature forest is scarce, and land is farmed again before reforestation is complete. The length of the cropping and fallow periods depends on a number of factors, most notably on the rate of decrease of yields and/or the increase in labor input necessary to maintain yields, and on the alternatives available to the farmer.

In the Grijalva River Valley, where Zinacantecos farm, there are many types of land. These range from rocky hillsides that rapidly decrease in fertility and require long fallow periods to flatlands near the river or its tributaries that repay the effort necessary to prevent the encroachment of weeds and may be farmed continuously. For the most part, the flatland is farmed by its Ladino owners with plow technology, and Indian renters are confined to the rocky areas and hillsides. An occasional Zinacanteco uses a plow with mules to farm flatland that he has managed to rent; and some hire Ladinos to plow for them with ox teams. However, I know of no Zinacanteco who has depended on plow technology for all of his fields in a given year, nor of any who has used plow technology for more than three or four years before changes in the landowner's plans took the land away from him. It is fair to say that all Zinacanteco farmers are involved in slash and burn agriculture; and that although all would prefer mature forest, most are working older land.

Preparing the Land

To prepare the land for seeding, the farmer must cut trees and brush if the land has been fallow for some time or gather stalks and weeds if the land was farmed the previous year. In either case he burns the accumulated vegetation at the height of the dry season (April or early May). If a farmer has new (i.e. reforested) land he usually cuts the trees and underbrush in January or February, so that they will be thoroughly dried by the time he is ready to burn. Axes, billhooks, and machetes are used for this work (see Photo 12). In an essay on corn farming, my informant José wrote. "It may take up to six weeks of work just for cutting down the trees on one tablon of land so that one almud of seed may be planted." But in most cases the land available to Zinacantecos does not require this much work, and other informants give lower estimates.

The four reports of actual labor input from informants who had recently cleared fields they described as including big trees are 12, 14, 20, and 24 man-days per tablon. The lowest figure is for land in an area that has been intensively farmed for many years (Zone 2), and it may properly represent the typical effort for the most fully reforested land available in that area. The land that required 14 man-days per tablon is shown in Photo 9. The remains of large trees are clearly present, as well as extensive limestone outcropping that no doubt prevented denser growth from taking hold. All in all, a figure of 20 man-days per tablon seems most appropriate for the cutting and burning of fully regrown land.[1]

The work of gathering weeds and stalks left on a field that was worked the previous year is done in April just before the burning, for the plants are then dead and dry. Informants' estimates for this job cluster around two and three man-days per tablon. Reports of actual labor inputs for preparation of fields that had been

[1] The figure of 10–14 man-days per tablon given in Cancian 1965a was based on data gathered in 1960–62 when more Zinacantecos worked in the area (Zone 2) that produced the 12 man-day report given here. In 1966 Zinacantecos had access to more fully reforested land.

9. Productive land that had large trees before it was cleared. It had been farmed for three years when the photo was taken, and was abandoned after a mediocre crop in the fourth year.

10. Zinacantecos inspecting land that has lain fallow for two years.

11. A group of hired workers weeding a lowland field. Most of them are not Zinacantecos.

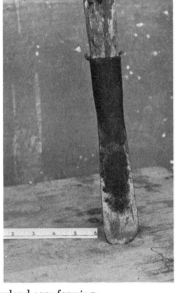

12. *Left*: Hoe, ax, billhook, and machete are the tools of lowland corn farming; sprayers like the one shown are a recent introduction. 13. *Right*: the metal tip of a dibble, used by Zinacantecos for planting corn; the entire dibble is five feet long.

cropped in the previous year are as follows: 2, 2.5, and 4 man-days per tablon in cases where the informant gathered stalks and weeds well before burning them; 3 and 4 man-days in cases where these two steps were done as one; and 6.4 and 10 man-days in two cases where the situation surrounding the work is not clear. The informant who reported a labor input of two man-days per tablon regarded this as very low, and explained that the job had been easy because the owner of the land had let his cattle loose to feed on the cornstalks after the previous year's harvest. A figure of 4 man-days per tablon, including time for burning, seems most appropriate as an average.

Burning, on either kind of land, is usually done at the height of the dry season, although it may begin early if the brush is dry enough. Some farmers postpone burning until May, saving themselves a trip by burning just before they seed. Burning is always done in one day, whatever the size of the field, although a second day is occasionally spent in making sure that the smoldering remains of the fire will not spread to adjacent fields or forests. Farmers with adjoining fields usually cooperate in the burning, thus permitting better control of the fire and minimizing the length of the firebreak that each must open and tend.

Seeding

Seeding is done in May, except where the rains begin in April (around Zone 9). The farmer's goal is to time his planting so that the seeds will germinate and sprout just as the rains begin. This gives the corn a head start on the weeds. If the seed goes in too early, it may sprout and dry before the rains or be uncovered and eaten by ants and rodents; if it goes in too late, weeds may hamper the corn. About half the time, it seems, some seed is lost and partial reseeding is necessary.

The seed is planted with a metal-tipped dibble (see Photo 13); and a Zinacanteco farmer usually puts six grains in each hole, leaving about one meter between holes in each direction. Agronomists recommend three or four grains to the hole and closer spacing, and many farmers do this when planting hybrid seed. Labor input for seeding is consistently reported at three man-days per almud; in fact, a man's workday is often regarded as done when he has seeded one-third of an almud. (A number of informants noted that seeding after the first rain is easier, since the ground is softer, and one gave estimates of two man-days for soft ground and three man-days for dry, hard ground.) Of the six men reporting on actual labor input, one gave a figure of 2.5 man-days per almud and the rest gave the three-day figure. Three said that no reseeding was necessary on their fields, and the other three reported 0.5, 0.5, and 1.5 man-days per almud for this task. This distribution of reseeding effort is consistent with estimates given by others.

Weeding

For most farmers, weeding is the biggest job of the agricultural cycle. And it is the step that limits the amount a man can seed, since the first weeding must be completed early in the season, before weed competition hurts the corn. Weeding is normally done with a hoe (see Photo 12); but when the land is extremely rocky, or when there are relatively few weeds, it may be done with a machete or billhook. In recent years some Zinacantecos have used chemical weed killers. But this practice has spread rather slowly, apparently because the farmers' lack of technical knowledge led to some colossal failures at the outset which frightened most farmers.[2]

---

[2] The weed killer used is a 2-4-D specific for broadleaf plants. It is effective in recently cleared fields ("new" land), where broadleaf weeds rather than grasses are the major problem; but it cannot be used where beans are interplanted with the corn. In the early 1960's a good many farmers tried chemical weed killer, but many of them lost their crops while waiting for the weed killer to work on old fields where grasses were the principal competition for the corn. Since then, knowledge of the limitations of weed killers has spread, in part because of informative commercials on the increasing number of radios in Zinacantan (see Chapter 7).
    The cost of using chemical sprays is fairly easy to calculate. One farmer who bought a hand sprayer (see Photo 12) in 1965 paid $280 and used $30 worth of chemicals per almud of seed. If water with which to mix the chemicals can be had near the fields, two men can easily spray an almud per day. One owner contracts to do the work of spraying, including all costs, for $60 per almud. This is, of course, much lower than the cost of weeding with a hoe on most fields; but it is close to the cost of hoe weeding on the recently cut fields where weed killers are most effective. If perfectly applied, weed killer could produce substantial savings for the limited number of Zinacanteco farmers in optimal situations.

The labor input for weeding varies tremendously. Weed competition increases with each year that a plot is used; and in the first few years it is inversely proportional to the length of time the land has lain fallow. After the first three or four years, even land that had been fully reforested needs two complete weedings, one in June and one in July. Moreover, the work must be done at the optimal time; otherwise it takes longer, and the crop will be damaged. My informant José, who prefers new land and who briefly used plow technology, exaggerates a bit in his essay on farming old land with the traditional hoe technology. However, he emphasizes the need to weed at the proper time.

On old land, when you begin, it takes 15 days to weed a tablon; later it takes 18 days, because the weeds are thicker; and in the end it takes 24 days to weed one tablon where one almud is seeded. But by this time the weeds have hurt the corn, the stalks are thin, and if a little wind comes all the corn will be blown over and lost. And if it doesn't rain for a couple of weeks, the corn is quickly hurt by the dryness. The land becomes very hard, and the corn dries; and on a hillside, corn can't stand the heat at all, for the land holds no moisture. There must always be two weedings; if not, the yield will be nothing. The second weeding takes two weeks [12 days]. So just using a hoe it takes altogether four weeks [24 days] of weeding for each tablon. But that's only if there aren't many weeds. When weeds are thick it takes up to six weeks [36 days], or even eight weeks [48 days].

For the most part informants report labor input that is lower than José's figures, even when they are on old land; but my notes include two reports of weeding that took 34 and 36 days per tablon respectively. In the first case, the informant considered his labor input normal. In the second case, the informant regarded his labor input as very high, and explained it by the fact that he was not able to begin weeding until very late in the season.

Table 4.1 displays the reports of labor input for weeding. I have divided the area farmed by Zinacantecos into two parts: an old area around Acala, and a new area around Venustiano Carranza. Fully regrown forest is common in the new areas, but in the old area only one of the seven cases involved land that had been fallow long enough to allow large trees to grow. Zinacantecos know that a farmer is more likely to find fully regrown forests (and hence better land) in the more distant zones.

TABLE 4.1
Reports of Actual Labor Inputs on Weeding

Years since last fallow period	Size of trees at end of last fallow period	Number of weedings	Labor input in man-days per tablon	Informant
Old area (Zones 1 and 2)				
1	small	1	8	A
1	—[a]	2	12	B
2	small	2	16	A
3	small	2	16	A
3	large	2	27	C
4	small	2	21	D
7	small	2	30	E
many	?	2	21	F
many	?	2	14	B
New area (Zones 6 to 9)				
1	large	1	3	F
1	large	1	4	G
2	large	1	6	F
2	large	1	6	H
3	large	1	6	F
3	large	1	6	H
4	large	1	12	I
many	?	2	14	G
many	?	2	34	J

[a] No substantial trees had grown during the two years this land had lain fallow.

For purposes of generalization, I will set labor inputs for weeding at six man-days per tablon for the first three years in the new area and at twenty man-days for all but the first one or two years in the old area. (In the old area, the labor input during the first three years might actually be 8, 16, and 20 man-days respectively.) I have not estimated labor input after the first three years in the new area, since my informants have only farmed this land for a short time.

Doubling

Doubling, which is done in September or October, involves snapping the cornstalk just above its midpoint, leaving the top of the plant and the ears pointing downward. This cuts off the food supply to the ears, and also helps protect them from rain while they dry out (Photo 23, p. 186, shows doubled corn). In addition, doubling minimizes the chance that a mature plant will be blown over by the wind. Occasionally, when a farmer has a very rocky field, he will not double, since the doubled plant would touch the large stones in the field. Some Zinacantecos who plant hybrid corn say that it does not need to be doubled, for it is shorter than the local strains and thus not so vulnerable to wind damage. My eight reports of actual labor input for doubling are 0, 1.5, 1.5, 2, 3, 3, 3.3 and 4 man-days per tablon. The informant who did not double had seeded hybrid corn on rocky soil. The other seven reports average 2.6 man-days. A standard figure of 3 man-days per tablon will be used for the doubling part of the production process.

Harvesting

The harvest involves two distinct steps: gathering the ears of corn, and removing the grain from the ears. It takes place between late October and January, the exact dates varying according to the condition of the fields and the farmer's schedule. The husk is opened at the end with a bone or metal implement about the size and shape of a builder's pencil, and the husk is separated from the ear with a two-handed spreading motion (see Photo 14). Then the ear is snapped off and placed in a bag. When the worker has filled a bag (about 100 liters of volume) he returns to his group's campsite and adds the corn to a pile there. Hired workers usually harvest on a piecework basis, and four bags full are considered a day's work. Very occasionally, when the camp is very far from the field or when the terrain is rough and the crop is poor so that it is hard to find ears, workers will demand to be paid by the day. Under these conditions they usually gather less than the standard four bags per day. More often, workers who are on piecework find that they can gather more than four bags a day if they work hard, and they are paid a quarter of a "day's" pay for each additional bag gathered.

Removing the corn from the ears is also done on a piecework basis; and a worker is expected to thresh eight bags per day. This commonly yields 24 almuds of grain. The actual process involves beating the dry ears with poles so that the grain falls off. For small amounts of corn, the ears are sometimes put in a net bag and the corn is allowed to scatter as it is hit. More often, a threshing platform is built, and a perforated cowhide is stretched horizontally so that the loosened grain falls through the holes and the ears stay above (see Photos 15 and 16).

The harvest is both the simplest and the most difficult step to describe in general terms. It is simple because the principal activities are commonly done on a piecework basis. On the other hand, one must collect scattered grains, choose seed corn, deliver the rent to the landowner, and transport the crop to the highlands; all these things occupy the time of the farmer, if not the time of the workers he may have hired. Since these activities may be interpreted either as part of the production process or as part of the auxiliary processes described in Chapters 3 and 5, I have excluded them from the present chapter. This is especially appropriate, I think, because some of them involve economies of scale that are not present in the production process itself. The standard figure for the harvest may be set at 1.5 man-days for each 12 almuds harvested (8 almuds per man-day), with the understanding that other work done at this point in the agricultural cycle will be covered by the inputs treated elsewhere.

Related Activities

I have now described all the steps needed to produce corn. However, almost every farmer invests a large part of his time in other activities: securing land, building a shelter, fencing, checking the progress of the crop in mid-season, paying rent, arranging transportation at harvest time, and hiring workers. Some of these tasks will figure in my calculation of the economic outcomes of alternative farming strategies (Chapter 6), but most will not. Here, I will briefly describe what is involved in each activity and then discuss the problem of estimating the time invested.

Group leaders who are seeking new land sometimes spend days wandering from ranch to ranch in the lowlands, but they are usually

14. Opening the cornhusk to remove the ear.

15. Grain is removed by beating the ears with poles.

16. The threshing platform is the most efficient arrangement, except for very small crops.

able to make tentative arrangements after the previous harvest and confirm them in February or March. The typical group member has only to visit the group leader and present him with a small gift of liquor to make sure of receiving land. Farmers whose group has moved to a new location often see their land for the first time when they arrive to prepare it for seeding.

The shelter built in the lowlands is usually shared by a single farming group. Usually, it is a low, grass-roofed structure that provides a place to sleep and protects food, extra clothes, and tools from the rains (see Photo 5, p. 35). Shelters are readily built from materials available on or near the campsite, and they generally last for more than one year.

Most landowners have cattle, and the farmers must protect their fields from damage by roaming stock. Arrangements for fencing vary from owner to owner, but typically the owner provides barbed wire and pays the farmers to cut posts and erect fences. (That is, the fence is treated as a permanent improvement to the landowner's property.) However, the wage is usually the lowest one current in the area, and the owner does not provide food. Thus the cost of fence-building to the farmer has two parts: a limitation on how he can spend his time; and the food required to feed himself or a hired worker. In succeeding years, of course, only minimal time need be spent to maintain the fences.

When the weeding is over in July, the fields require no more work until the cornstalks are doubled in September. However, most Zinacantecos make a trip to the lowlands in August to check on the progress of the crop. If they go by truck or happen to have fields very near their homes, the trip may be made in one day; but two days is the usual time spent.

After the corn has been harvested and threshed, a number of small tasks remain, which are usually handled by the farmer or young sons who work with him. The ears that have been selected for next year's seed must be degrained. Bags must be carefully sealed for shipping. Kernels that have been scattered around the threshing area must be gathered up. The rent, which is almost universally paid in corn, must be taken to the owner's house or some other central place on the ranch. And arrangements must be made for transporting the crop to market or to the highlands.

As we saw in Chapter 3, the overwhelming majority of Zinacantecos hire workers to help them in the fields. Usually, these are one or two men from Chamula or one of the other Indian townships surrounding San Cristóbal, and they can be found and hired in a few hours during a trip to the city. A few farmers hire Zinacantecos from their own hamlet. Large operators who need many workers may have some trouble finding them. However, large operators provide more consistent work, since they need help at every point in the agricultural cycle; and they often have continuing arrangements with the same workers.

Estimating the time required for all these chores is extraordinarily difficult. Since the time involved is mostly that of the farmer himself, such an estimate will not influence my calculation of the money invested in workers. Most of the time involved, in fact, is time in which the farmer might have little else to do. Nevertheless, these subsidiary activities do represent a real investment of effort and are necessary to the overall success of a corn-farming operation, so some time estimate, however arbitrary, should be made.

I will assume that every farmer must spend one day each year in finding land to rent. Farmers who are changing locations probably take longer, and group leaders obviously do. (But the greater investment of time by group leaders may be more than balanced by the advantages they have in selecting their plots.) The time necessary to build a shelter and fence the fields is substantial only when a farmer rents reforested land, and even then the time spent fencing is paid for by the landowner. The standard input for preparing new land is 20 man-days per tablon, and I will assume that uncompensated time devoted to fencing and time devoted to building or repairing a shelter are both absorbed in this figure. Finally, as we have seen, every farmer spends two days (or one day plus carfare) to inspect his fields in August.

Large-scale operations have some advantages in these tasks. The time spent finding land and inspecting the fields in August remains the same in a large operation; and the time investment in transportation and marketing (see Chapter 5) does not increase in proportion with the size of an operation. These two factors are direct savings of scale. A more important advantage exists when a large operator's income per unit of production is greater than his costs for workers, transportation, and marketing; for in this case he has proportionally less of his own time invested in each unit of production. However, some costs accompany these advantages. In the section that follows I will establish rules for estimating the additional time that larger operators must devote to "administrative" activities during the harvest period. The number of days so spent increases with the size of a farming operation, and it is necessary to have some idea of the costs of operating on a larger scale.

The Use of Workers

The hiring of workers to help with lowland fields may vary greatly with individual circumstances. For example, one of the farmers described in Appendix B, Marian, makes relatively short trips to his fields and hires a relatively large number of workers. Part of this is explained by his need to finish at one of the locations that he farms so that he may get to the other; but he also dislikes going to the lowlands unaccompanied. Romin is unusual in another way: in 1966 he hired workers to do all his harvesting while he remained in the highlands working at another job. In establishing standards for hiring workers, I have excluded these idiosyncratic factors from consideration. My estimates are thus unrealistic for a small proportion of the population, but on the whole they are a good approximation of the actual pattern of constraints under which Zinacanteco farmers work.

Table 4.2 shows the labor input required for one tablon of land with one almud seeded. The first column of the table simply summarizes the earlier discussion in this chapter. The second column, which states how much time a farmer will put in himself before hiring workers, is to some extent arbitrary. For seeding and weeding, the figures shown are supported both by my experience with actual cases and by the fact that these tasks must always be completed within a short period, whatever a farmer's inclination or par-

TABLE 4.2
Labor Input and Hiring Workers

Operation[a]	Labor input per tablon (man-days)	Time farmer will work on his own (days)	Tablon (almud) units he can work alone
Prepare new land[b]	20	20	1
Prepare old land	4	8	2
Seed	3	6	2
Weed new land[c]	6	12	2
Weed old land, 1st	12	12	1
Weed old land, 2d	8	8	1
Doubling	3	6	2

[a] Figures for harvests are discussed in the text.
[b] For the first year on reforested (new) land only.
[c] For the first three years on new land only.

ticular situation. The figures for the other tasks are also consistent with the behavior of farmers as I know it, but there are no technological constraints to keep a given farmer from increasing his own labor input on these tasks in order to avoid hiring workers. In general, the figures given will hold for operations where the amount seeded is two almuds or a simple multiple of two almuds, since very few Zinacantecos would work twice the periods shown without hiring workers. It is quite possible for a farmer to seed slightly more than two almuds and do the extra work himself or with the help he can get from his sons.

The problems of estimating harvest time are more complex, and so is the behavior of farmers. A farmer may come with his workers, do the whole job, and go home; or he may have workers help him with the big jobs and then spend many days alone finishing up odds and ends. (Antonio's harvest schedule in Appendix B is an example of the second approach.) And a large operator is likely to let his workers do the straightforward harvesting tasks while he does related chores that require his judgment. Although it is realistic to assume that a farmer always works along with his employees before the harvest, this assumption is inappropriate for work during the harvest period. The problem is not how much labor is needed for harvesting but how much the farmer himself will contribute, how many workers he will hire, and how much of his time will go to tasks not directly related to getting in the crop. Various sets of assumptions could produce realistic estimates, and I have chosen to assume the following. An average farmer will spend up to 15 days working without help. If the harvest work takes more time than this, he will hire workers.[3] All but the last worker hired will also work fifteen days; and for every worker after the second hired, the farmer will spend a day on activities that do not contribute directly to getting in the harvest. For example, if 45 man-days of work are needed, the farmer and two workers will each do 15 days. With 46 man-days, the farmer will do 14 and three workers will do 15, 15, and 2. And with 90 man-days, the farmer will work 11, and six workers will do a total of 79. These standards will be used in Chapter 6.

Additional harvest expenses are incurred by a farmer whose crop is exceptionally poor, since workers will not do piecework if the ears of corn are small or widely scattered. A realistic standard for these extra inputs is probably 10 percent of the total harvest input for each fanega (12 almuds) less than six fanegas' yield per almud seeded. Thus a yield of five fanegas adds 10 percent to costs, and one of three fanegas adds 30 percent.

Yields

Yields vary greatly from zone to zone. The main factor affecting the yield is soil fertility, but annual climatic fluctuation and the peculiarities of each farmer's land are also important. As Table 4.3 (in combination with the maps) shows, the land closest to population centers typically gives lower yields, presumably because it has been exploited intensively over the years and has not been allowed to go through full reforestation cycles.

Zinacantecos who rent in each of the zones naturally base their decisions about renting on the characteristics of the specific piece of land they are considering. They are, however, quite aware of the overall trends. Since the land that gives higher yields is usually

---

[3] In building models in Chapter 6, I have treated part days as whole days worked by the farmer, e.g., a crop that requires 22.5 man-days of work is treated as 23 days, 15 of which are worked by the farmer.

TABLE 4.3
Yields by Zone, 1957–66

Category	Zone						
	1	2	3	4	5	6	7–9
Mean yield in almuds per almud seeded	56	92	72	107	137	124	157
Range of yields	24– 108	36– 180	60– 84	24– 180	72– 288	48– 288	120– 216
Number of farmer-crop-years	12	71	2	14	17	22	5
Number of informants[a]	6	12	2	6	5	6	4

NOTE: These figures are based on the work histories. Informants reported on 151 farmer-crop-years. Eight farmer-crop-years were in areas to the west of Zone 1 and were not counted, leaving 143 farmer-crop-years on which this table is based.
 [a] Twenty different informants are represented.

farther from Zinacantan than the land that gives lower yields, they must weigh a number of factors. Differences in labor input between zones were discussed above, and in Chapter 5 I will describe the substantial differences in transportation costs and marketing opportunities. In this section, I will try to give reliable estimates of yields that may be used to build an overall characterization of farming in each of four zones. (The text and the tables use the standard almud unit, although Zinacantecos normally discuss harvests in fanega units.)

The distribution of the yields on which Table 4.3 is based is displayed in Table 4.4. Although the distribution as a whole is very close to a normal one, the zones for which few cases are available show ambiguous patterns; and there are simply not enough cases to permit setting standards for each zone directly in terms of these figures.[4] In setting the standard yields shown in Table 4.5, there-

[4] For Zone 3 there are only two cases. However, I collected a number of cases from Zone 3 for the period before 1957, and other less concrete reports on the zone indicate that the standard yield would probably be close to 120 almuds per almud seeded. Zones 4 and 5 lie on a continuum between Zones 2 and 6 for most purposes. On similar criteria, Zones 7 and 8 may be lumped with Zone 9.

TABLE 4.4
Informants' Reports of Yields

Yield in almuds per almud seeded	Responses, by zone									Total no. of cases
	1	2	3	4	5	6	7	8	9	
12										—
24	2			1						3
36	3	1								4
48	2	3		1		1				7
60		5	1	2						8
72	3	10			1	2				16
84		10	1	1		2				14
96	1	16		2	2	2				23
108	1	12		2	2	2				19
120		11		2		4	1	1		19
132		1			1	2				4
144				1	4	3				8
156				1	2	1	1			5
168		1		1	1					3
180		1		2	1	1			1	6
192										—
204										—
216						1			1	2
228										—
240					1	1				2
TOTAL	12	71	2	14	17	22	2	1	2	143

NOTE: These figures are rounded off to the nearest fanega (one fanega = 12 almuds). The means in Table 4.3 are based on the figures before rounding.

fore, I have had to use information that was not systematically gathered and cannot be presented in tabular form.

First note that Table 4.5 lists four yields for each zone: a normal yield; a good yield, 25 percent above normal; a poor yield, 25 percent below normal; and a disaster yield, which is 50 percent of the normal yield. A comparison of the standards (Table 4.5) with the actual informants' reports of yields (Table 4.4) shows that these four yields represent the range of actual yields fairly well (with the exception of Zone 9, which is discussed below), and that the standard disaster yields and poor yields occur more frequently in the nearby zones. For example, the normal yield for Zone 1 is set at 72 almuds; but the actual mean (Table 4.3) is 56 almuds, and seven of the twelve cases shown in Table 4.4 report yields of

TABLE 4.5
Standard Yields by Zone
(*almuds per almud seeded*)

Type of yield	Zone			
	1	2	6	9
Normal	72	96	120	144
Good	90	120	150	180
Poor	54	72	90	108
Disastrous	36	48	60	72

less than 60 almuds. The two Zone 9 figures in Table 4.4, by contrast, are at the good yield level and above.

I have set my standard yields as they are in order to produce a closer approximation of Zinacanteco thinking about the various zones. No Zinacanteco would farm a lowland plot if he knew for sure that his yield would be less than 72 almuds per almud seeded. The fact that the yields in Zone 1 tend to be less than this minimum standard is consistent with the fact that very few Zinacantecos actually do farm in this zone. A few try it because the land is conveniently located, especially for the farmer who is running a small operation; but most of these do not stay long (see Marian's case in Appendix B). The normal yield in Zone 9 is set at 144 almuds, because that is, in my experience, a figure that any Zinacanteco farmer would be delighted to accept as a guaranteed yield—that is, he would rather have it than take even odds that it might go higher or lower. In the long run the mean yield in Zone 9 may be even higher, though not by much.

The information on Zone 9 that I used to supplement the two case reports shown in Table 4.4 is typical of the kind of information I have in addition to that shown in the table. I discussed yields in Zone 9 with perhaps ten farmers. Two were a group leader and a member of his group who farmed there in 1966; they reported their own yields, discussed the yields of other group members, and described their conversations with other farmers working in the area. Another informant was a man who planned to farm in Zone 9 in 1967 and had investigated the situation for his own purposes. Another had a brother farming there in 1966. And another farmed in Zone 7 but knew farmers who worked in Zone 9. I also talked with a Zinacanteco truck driver who brought harvests in from the area and was considering farming there himself, using hired workers to do all of his work.

On the basis of these conversations I formed an impression of what farmers expect as yields, what would please them, and what would disappoint them. These three figures are given as the normal, good, and poor yields respectively. No one I talked with reported or had even heard of a yield lower than the poor yield of 108 almuds, though some had heard of yields higher than the good yield of 180 almuds. A very bad year, I think, would produce something very close to the disaster yield of 72 almuds. Even though no disaster yield has actually appeared during the few years that Zinacantecos have been farming in Zone 9, I feel that the farmers do consider it possible in Zone 9, and I have therefore included it in the figures.

The standards shown in Table 4.5 represent the range of results with which a farmer must be willing to deal. In each case, the normal yield is intended to represent the modal yield, the yield on which the farmer might be expected to base his farming strategy. The good and poor yields are intended to be results that occur often enough to be easily accepted by the farmer. And the disaster yield is one for which the farmer might be expected to make some provision, although the circumstances bringing it about would always be considered extraordinary. For Zones 2 and 6, the figures in Table 4.5 fit the empirical distribution of yields quite closely. The fact that Zone 1 often shows yields at the disastrous level is well known to Zinacantecos and, as I mentioned above, they avoid farming there despite the other advantages. All my experience points to the conclusion that in 1966 a great number of Zinacantecos saw the prospects for Zone 9 in the terms in which I describe them.

Variance in Labor Input and Yields

The labor input required by lowland farming varies greatly with the land. Table 4.2 displays standard estimates of the labor input required for each step in the agricultural cycle, both for recently cut forest and for land that has been continuously worked for several years. The same table's estimates of how long a farmer is usually willing to work at each point of the agricultural cycle may be used in combination with information on scale of operation to determine how many workers farmers are likely to hire. Yields also vary greatly with the quality of the land; and, on the whole, the land varies from zone to zone, with the nearer zones giving poorer crops. Table 4.5 displays standard estimates for the range of yields in four zones spread across the range of conditions faced by Zinacantecos in the lowlands.

FIVE

Transportation and Marketing

ZINACANTECOS farm across the entire Grijalva River Valley, from
Chiapa de Corzo in the west to Porvenir in the east; and a few find
their land to the west of Chiapa de Corzo and Tuxtla Gutiérrez.
As the maps clearly show, distance from Zinacantan and access to
roads vary greatly within this area. The cost in time and money of
getting to and from the fields varies correspondingly, as does the
cost of transporting the crop at harvest time. No matter where they
farm, most farmers have a choice of markets for their product.
This choice ranges from immediate bulk sale at government-spon-
sored receiving centers immediately after harvest to personal retailing of small lots at the peak market prices, which usually come
eight to ten months later.

Transportation and marketing are important economic aspects
of Zinacanteco farming, and they are easily separated from the
organizational features described in Chapter 3 and the production
process described in Chapter 4. I have chosen to deal with them
together in this chapter because they are both auxiliary to the production process—and, more important, because they are the economic features of Zinacanteco farming that have changed most
profoundly in the last ten years. In order to set the stage for the
economic analysis of the choices made by Zinacanteco farmers
(Chapter 6), and in order to present ethnographic detail in the
least confusing manner, I will give a synchronic description of
transportation and marketing in the 1966 ethnographic present. In

1966, trucks were the predominant means of transportation for farmers whose homes were near roads, and 50 percent of Nachig farmers sold some of their crop at the new government receiving centers. The reader should remember, however, that only ten years before there was virtually no truck transportation and no opportunity to sell to government receiving centers. These and other changes in the period 1957–66 will be the principal concern of Chapter 7.

I will first describe the road system and the cost of transporting both people and corn. In these sections the text is intended to supplement the data presented in the maps and tables. The later sections of the chapter will deal with the marketing alternatives available to Zinacanteco farmers.

Roads

The major roads in the area in 1966 are shown on my maps. The Pan-American Highway is a wide, two-lane asphalt road. The all-weather roads are fair to excellent gravel roads that cross all rivers and streams on modern bridges. (Although a short stretch between Flores Magón and Venustiano Carranza was still a bit of a challenge in the 1966 wet season, even it was passable at all times by any type of vehicle.) The dry-season roads vary greatly. The road between Pujiltic and Venustiano Carranza, for example, includes a ford across the Río Blanco and can be difficult for a sedan even in dry season. Other dry-season roads are simply paths across open fields. Besides those shown on the map, there are hundreds of local dry-season roads that are used at harvesttime.

My log for a trip I made in the dry season (December) from Porvenir to San Cristóbal, a total distance of 65 miles, illustrates the relative condition of the roads. The 14 miles from Porvenir to Pujiltic took 75 minutes (11 m.p.h.). The 21 miles from Pujiltic to the Pan-American Highway took 65 minutes (20 m.p.h.). And the largely uphill trip along the Pan-American Highway to San Cristóbal took 45 minutes (40 m.p.h.). The distance from San Cristóbal to Acala, around the "elbow" near Chiapa de Corzo, is roughly the same (63 miles).

Travel Time and Transport Fees

Zinacantecos going to the lowlands either walk down the escarpment shown in Photo 4 (p. 18) or travel by truck along the roads. When they walk, they often use mules to carry their food and tools.[1] When they go by truck, they normally hail passing vehicles that travel the major roads anticipating this kind of trade; and their trip may involve one or more changes of vehicles at crossroads. Occasionally, a large work group finds it economical and convenient to hire one truck for the entire trip. In transporting the harvest, specific arrangements for a caravan of mules or a truck are usually made. The means and ranges for walking time, carfare, and transport of corn by mule and truck between Nachig and the various zones are given in Table 5.1.[2] Since there is no motor road to Apas, Table 5.2 gives only the figures for walking time and mule transport.

In 1966 the majority of Nachig farmers hired Ladino truckers to carry their harvest; but many Nachig farmers and most Apas farmers used mules. As illustrated by Marian's case in Appendix B, the hiring of mules from fellow Zinacantecos involves complications and costs beyond the set cash price given in Tables 5.1 and 5.2. In a few places mule transport is still more practical than truck transport, even for Nachig farmers. For example, two of the seventeen Zone 2 locations farmed by Nachig men cannot be reached by truck, and two others are so close to Nachig that mule transport is significantly cheaper than hiring trucks. Of the remainder, seven are clearly less expensive by truck, and six are less expensive by mule only if the farmer does not figure the extra complications and

[1] Most pack animals used by Zinacantecos are mules. Some horses are also used. Burros are sometimes used by lowland residents for local transport;· but they are considered too small and weak for travel up and down the mountains, and are never used by Zinacantecos.

[2] These tables are based on reports by my principal informants, who gave comparable figures for all the locations mentioned by Nachig and Apas farmers in the survey. Their statements were checked with other informants. The cash rates for each type of transport are fairly conventional, and most of the difference within zones may be attributed to the distance between locations included within zone boundaries rather than to differences in rates charged from the same location.

TABLE 5.1

Travel Costs and Transport Fees Between Nachig and the Lowlands

| | | Means | | | Ranges | | |
| | Truck fare (one way) | Walking time (hours, one way) | Transport fee ($ per almud) | | Walking time (hours, one way) | Tranport fee ($ per almud) | |
Zone			Mule	Truck		Mule	Truck
1	$5.00	—	—	$.55	—	—	$.55
2	—	8	$.95	.95	6–10	$.65–1.25	.90–1.10
4	—	10	1.20	1.10	9–10	1.00–1.25	1.10
5	—	12	1.25	1.15	11–13	1.25	1.10–1.25
6[a]	—	15	1.65	1.40	14–16	1.65	1.35–1.65
7	6.00[b]	—	—	1.00	—	—	.90–1.10
8[a]	13.50	18	—	1.35[c]	18	—	1.35
9	9.60	—	—	1.10	—	—	1.10

NOTE: Figures are rounded to the nearest hour and the nearest $.05. Nachig men do not farm in Zone 3. A dash indicates that Nachig men seldom if ever walk/ride to that zone, or that mule transport is not used from that zone.

[a] Walking to these zones requires two days, and food for the trip is charged at $2.50 per one-way trip in my calculations.

[b] Most ranches are two hours' walk beyond the usual truck stop.

[c] This is the charge by road to Nachig; but most locations are far from the road and require another $.90/almud for oxcart transport to the road.

costs of mule transport. For Zones 4 and 5, truck transport to Nachig is slightly less expensive than mule transport. Zone 6 is the limit for the practical use of mules, and it was the limit of Zinacanteco farming before truck transport became common.

If a farmer decides to sell his corn in the lowlands rather than taking it home, the transportation costs shown in Tables 5.1 and 5.2 do not apply. Government receiving centers are distributed throughout the area farmed by Zinacantecos; and, although the distance from a farmer's fields to a receiving center may vary greatly, it is always relatively short compared to the trip home. Since a large part of the charge made by truckers seems to be assessed for the time they must wait while the corn is processed at a receiving center, the variance in charges is not as great as the

TABLE 5.2
Travel Time and Transport Fees Between
Apas and the Lowlands

	Means		Ranges	
Zone	Walking time (hours, one way)	Mule transport ($ per almud)	Walking time (hours, one way)	Mule transport ($ per almud)
2	6	$.80	2–6	$.50–.90
3	9[a]	1.65[b]	7–13[a]	1.50–1.90
6	14	—[c]	14	—[c]

NOTE: Figures are rounded to the nearest hour and the nearest $.05. Apas men farm in Zones 2, 3, and 6 only.

[a] In addition, it takes about one-half hour and $.40 to $.50 in fare to cross the river in dugout canoes operated by Acala people.

[b] In addition, it costs about $.07 per almud to get the corn across the river. Part of the cost shown in the table is for oxcarts hired to take the corn to the river.

[c] Usually, the crop is sold near the fields. In extraordinary cases, corn may be brought to Navenchauc by truck and taken to Apas by mule, but this is very expensive.

variance in distance from field to receiving center. An estimate of $.50 per almud is probably close to the typical rate. The many farmers who sell part of their crop directly to landowners or truckers pay nothing for transport, for such sales are usually completed in the fields. Details on this type of sale are given below.

Corn Taken Home and Corn Sold

Every farmer must end the agricultural cycle with a supply of corn in his home—enough to feed his family until the next crop is in, and enough to pay the workers who will help him during the coming year. A few farmers get some of their food and working capital from highland fields and from the closer of two lowland locations, thereby avoiding excessive transportation costs. Others sell all their corn in the lowlands just after harvest and pay workers with cash. But the majority of Zinacantecos follow what I will define as standard practice; and many farmers bring almost all their corn home, preferring to market it in San Cristóbal later on in the season. I will set the standard minimum taken home at 100 almuds, plus one additional almud for each worker-day used in production. The 100 almuds is the amount usually needed for the family's food

TABLE 5.3
Use of Marketing Outlets, 1966
(*from survey*)

| Town | Outlet | | | | | Mean number of outlets used per farmer |
	S.C. market	S.C. middle-man	Govt. rec. center	Private buyer in low-lands	Zinacan-teco specu-lator	
Nachig ($N = 191$)	88%	12%	49%	65%	24%	2.4
Apas ($N = 103$)	55	16	33	58	52	2.1

and for sale in the San Cristóbal market.[3] The extra almud per worker-day covers the pay and food of hired workers. The remainder of a farmer's crop may be regarded as commercial production, which he will market as best suits his situation.

Markets

After paying his rent and setting aside corn to feed his family and pay his workers until the next crop is in, the typical Zinacanteco farmer has substantial amounts of corn to market. The major marketing alternatives are four: sale to government receiving centers in the lowlands; sale to private buyers in the lowlands; retail sale in the San Cristóbal market; and sale to Zinacanteco middlemen who will resell the corn in the San Cristóbal market. The use of these alternatives by Nachig and Apas farmers is shown in Table 5.3. The table also lists the sale of corn futures (usually in small amounts) to Zinacanteco speculators who provide the farmers with cash before the harvest is in.

Most farmers sell corn through more than one outlet, as the table indicates. Nachig farmers, who have easier access to trans-

[3] The corn consumed by a family varies with family size, the appetite of individual members, and the availability of corn. Reports from five informants, listed from the most accurate (in my opinion) to the least accurate, are as follows: 73 almuds per year (a middle-aged couple); 134 almuds (a couple and five children), 78 almuds (a couple and three children), 73 almuds (a couple and three children), 73 almuds (a young couple and two adult women). The first couple listed are both large, stout people by Zinacanteco standards. The last couple listed are both fairly small and thin.

17. A government corn receiving center. A truck scale and a building where quality tests are made in the foreground.

18. Zinacantecos awaiting buyers for their corn in the San Cristóbal market.

19. A farmer and his mules taking rent from his fields to the landlord's house.

20. Trucks awaiting passengers and cargo in San Cristóbal.

portation than Apas farmers, use a greater number of outlets on the average; and they sell substantially more often in the San Cristóbal market and to government receiving centers and substantially less often to speculators in their own hamlet.

Sale to Receiving Centers and Private Buyers

The receiving centers are the key to marketing corn in the lowlands, even for those Zinacantecos who never go near them. The centers (see Photo 17) are part of a nationwide system operated by independent government agencies.[4] Since they buy at a standard price throughout the harvest season, they provide a stable base for the corn market, lowering the risk to truckers and other private buyers who purchase corn in the fields. Before the advent of the centers, lowland buyers were usually speculators who warehoused corn for resale at peak prices months later, but now many of them act as agents who bulk and transport corn to the receiving centers. They do this at a profit, but the profit is closer to an agent's fee than to the return a speculator might expect for a risky, long-term investment.

The farmer who wishes to sell part of his crop in the lowlands must consider many factors in deciding whether to take his corn to a receiving center or to a private buyer. Corn sold to the centers must meet certain quality standards and must be delivered in standard bags. The farmer must go to a bank (usually in Tuxtla Gutiérrez, Chiapa de Corzo, or San Cristóbal), pay a cash deposit for each bag he anticipates using, take the receipt for his deposit to a center and exchange it for bags, take the bags to his fields and fill them, and find transportation (usually a Ladino trucker) to get the corn to the center. At the center, his corn is weighed and tested for quality, and he is given a receipt for the sale.

During the peak seasons, the banks that handle the financial part of the operation send cashiers to the centers to pay farmers on the spot; but delays of at least a day between delivery of the corn and receipt of payment are common. The description in Ap-

[4] The two major agencies are ANDSA and CONASUPO. The development of the receiving centers is described in Chapter 7.

pendix B of Antun's sale to the Acala Center is typical in many ways. If a farmer is not bilingual and relatively sophisticated, as Antun is, he may be put off by the complications of dealing with the centers. However, many farmers get their corn to the centers through cooperative arrangements with members of their work groups who can handle the complications.

The centers buy by weight at the set price of $940 per metric ton. From this, they subtract various taxes, and also discount small amounts if the corn is not up to their quality standards. According to an official of the agency that operates the centers, the price actually received by farmers averages $880 per ton, and my inspection of receipts shown me by Ladino farmers confirms this estimate. In terms of the volume measures usually used by Zinacantecos (see Appendix C), this works out to $9.50 per almud—which is higher than the mean price in local markets but lower than the peak price.

Sale to a private buyer may be more attractive than sale to a receiving center if a farmer wants to sell only a small amount, if he is in a hurry, or if he lacks the skills or connections to get himself and his corn through the complications of selling to a center. Private buyers usually seek farmers out in the fields and provide the bags in which the corn is to be transported and stored. Since the purchase is made in the fields, farmers also avoid paying for transport. As shown in Table 5.4, private buyers in 1966 offered prices ranging from $7.50 to $9.00 per almud. I will set the stan-

TABLE 5.4
Price and Volume of Lowland Sales, 1966
(*from work histories*)

Buyer	Means		Ranges	
	Price per almud	Almuds sold	Price per almud	Almuds sold
Private (6 reports)	$8.35	36	$7.50–9.00	24–48
Receiving centers (4 reports)	9.50	360	9.50	276–540

dard price for sale to private buyers at $8.35 per almud, or $100 per fanega (the 12-almud fanega is customarily used to calculate trucking rates and bulk sales).

At $100 per fanega, the farmer is getting $14 less per fanega than he would get at a receiving center. Transport to a center typically costs $5 or $6 per fanega, so the farmer is actually paying $8 or $9 to the middleman for his services. Farmers selling small amounts may consider this procedure the best way to avoid dealing with banks, contracting for trucks, and so on; but when the amount sold is more than three or four fanegas (about one-half ton), sale through a middleman means a substantial loss in profit. The table shows a great contrast between the volume sold to private buyers and the volume sold to receiving centers by farmers reporting their 1966 activities. The dramatic contrast would almost certainly soften if the sample of farmers were larger. Nevertheless, it seems clear that Zinacanteco farmers make use of the two marketing alternatives available in the lowlands according to the volume they wish to sell. In Chapter 7 (especially Table 7.7) we will see that this clear distinction is a recent development.

Some farmers, of course, do not sell any of their crop in the lowlands, but take it all home and do their marketing as described in the following section. Table 5.5 shows the distribution of sale in the lowlands against the size of crops after rent is paid. As might be expected, farmers with a large crop are more apt to sell some

TABLE 5.5
Farmers Selling in the Lowlands
(*from work histories*)

Sold in lowlands	Size of crop after rent (almuds)			
	12–78	84–153	156–252	264–1080
Yes ($N = 40$)	2	8	11	19
No ($N = 38$)	8	17	7	6
TOTAL	10	25	18	25

NOTE: This table is based on work histories for the period 1962–66. In 1966 alone, 11 of 14 farmers (including the six with the largest crops after rent) sold in the lowlands. The classification of crop size is based on the entire work history data of 132 farmer-years from 1957 to 1966, and represents quartiles of that population.

of it in the lowlands immediately after the harvest, whereas those with a small crop are more apt to take the entire harvest home.

The San Cristóbal Market

Though the opening of receiving centers has increased lowland corn sales in recent years, the San Cristóbal market is still an important outlet for the corn sold by Zinacantecos. The marketplace is the area's major trading center for perishable foodstuffs.[5] Its main building and grounds encompass about a hundred permanently established stalls, which sell a wide variety of foods and household items. An even greater number of merchants set up temporary stalls on the terraced steps of the main market building and on the flat areas around the building and the permanent stalls (see Photo 18). Buses and trucks make the market their principal stopping place, and people regularly seek out business contacts and friends there. Activity is constant from early morning until midafternoon every day, and the busiest period is usually between 10 A.M. and noon.

The sale of corn in the marketplace is dominated by Zinacantecos. A few of the permanent Ladino vendors sell corn, as do Indians from other highland townships, but Zinacantecos control most of the corn outlets in the market itself. When prices are low and Zinacantecos are busy with their lowland fields, the number of sellers drops off radically, and occasionally there are none to be seen by early afternoon. But typically there are enough sellers to make a competitive market, and buyers go from one to another inspecting the corn and asking prices. Word gets around quickly, and the day's price is usually stable by the peak selling period. Table 5.6 summarizes observations of the number of sellers in the market on sample days over a five-month period.[6] The permanent

[5] There is a second but much smaller official market. The main market is described in more detail by Bahr 1961, Zubin 1963, and Capriata 1965.

[6] My assistant, who knew the market and the middle men well enough to make this count, had a regular job that took him out of town from time to time. Note the fluctuation around November 2, when the Fiesta of Todos Santos (All Saints' Day) is observed; in Zinacantan this celebration resembles a harvest festival. January 22 is the Fiesta of San Sebastian, the second most important of the year (see Cancian 1965a).

TABLE 5.6
Corn Sellers in the San Cristóbal Market, 1966–67

Date	Dealers	Producers	Date	Dealers	Producers
Oct. 1966			Nov. 1966		
1	9	11	6*	6	21
2*	11	14	7	4	18
3	9	18	8	3	13
4	10	13	9	3	16
5	9	15	10	5	12
6	11	13	Dec. 1966		
7	10	19	4*	5	11
8	11	17	5	4	9
9*	9	11	10	6	12
10	10	9	11*	5	11
29	11	27			
30*	12	29	Jan. 1967		
31	9	25	1*	6	10
Nov. 1966			15*	7	17
1	5	3	22*	4	7
2	2	—	Feb. 1967		
3	8	18	5*	8	14
4	10	15	17	9	16
5	9	21	26*	7	15

NOTE: All those listed are Zinacantecos. In addition, on November 9 two sons of one of the permanent stall owners set up in the open market. And Indians from Chamula appeared on three occasions: three on November 7, two on November 10, and one on February 17.
Asterisks indicate Sundays. There is no clear "market day" pattern in San Cristóbal.

stalls that sell corn as well as other products were not counted.

The price of corn in San Cristóbal varies with the seasons, presumably on the basis of supply and demand. Table 5.7 shows the mean price for each month in 1966. The peak in July–August and the dip around harvesttime is typical of price fluctuation in recent years. However, the mean price in 1966 was atypically low and the range of prices was atypically small. The price in 1965 had ranged from $8.30 to $11.35, with a mean of $9.40; and in 1967 the price went from a low of $8.85 to a spectacular high of $12.00 during a period when trading in the market was suspended because of price controls imposed by municipal authorities. The mean for 1967 was $10.45.

Many special circumstances enter into this price fluctuation over

TABLE 5.7
Corn Prices in the San Cristóbal Market, 1966
(*pesos per almud*)

Month	Mean price	Month	Mean price
January	$9.00	July	$9.45
February	8.70	August	9.30
March	8.70	September	8.85
April	9.00	October	8.25
May	9.00	November	8.55
June	9.15	December	9.00

NOTE: The mean prices are based on daily observations. Procedures used in handling the data are described in Chapter 7. The mean price for the year was $8.85, and the ratio of the highest price ($9.45) to the lowest ($8.25) was 1.15 : 1.

the years, and some of them cannot be usefully explained without further knowledge. A standard price for the synchronic analysis focused on 1966, given the variation in the period 1965–67, is probably best set at the mean of the three annual means involved. Thus I have set the standard price for 1966 at $9.60 per almud even though no monthly mean in 1966 reached that level. This procedure can only be justified in the present context by the fact that I have used the actual prices to get an estimate of what a Zinacanteco producer might expect from the market if he chooses to sell his corn in San Cristóbal. The variation from year to year suggests that the sellers must take into account the uncertainties of the market; and the highs in 1965 and 1967 suggest that the motivation to endure those uncertainties may come from the possibility of extraordinary returns in good years.

A farmer who wants to sell simply brings his corn to the marketplace, usually by truck, and pays a peso per nine-almud bag for the right to spread out his product in the area traditionally used by the corn sellers. He may or may not indulge in the luxury of hiring a market porter to carry his huge bag from the truck to the selling area. Though all the component costs are known, it is hard to set a standard figure for the expenses involved in selling corn in San Cristóbal. For transport to the market, a Nachig man will pay $1 for himself and $1 for a nine-almud bag of corn; and at the end of the day he must pay another $1 to return home. An Apas man must somehow get his corn to the road at Navenchauc, and even then he pays twice as much truck fare as a Nachig man. A market porter charges $.50 per nine-almud bag, and the market tax is $1 per bag of corn.

A farmer can expect to sell up to 12 almuds at retail in a normal day, depending on market conditions and on the price he asks. He may sell his corn quickly or have much of it left at the end of the day, but if he lowers his price enough he will always find a buyer somewhere in the market. Many Zinacantecos who need a bit of cash for a family necessity will bring in two or three almuds of corn, sell them quickly, and spend the rest of the day doing errands and enjoying the city. Perhaps the best that can be done is to assume that farmers whose main business is retailing corn will sell 12 almuds in a day and to figure the expenses of sale accordingly. (For those who sell less, fixed expenditures such as personal transportation are simply considered part of a day in the city.) On this basis, the standard cost of selling an almud (figuring a 50 percent probability of hiring a market porter) is $.40 for a Nachig man and $.85 for an Apas man (including a charge of $2 for transport from Apas to the highway in Navenchauc).

The farmer who wishes to free his time for errands in the city may sell his corn to one of the Zinacanteco middlemen who operate in the marketplace on a regular basis. These dealers are usually willing to pay the going market price, or very close to it, and they often pressure farmers who are retailing their own corn to sell to them in bulk. The dealers have better connections in the marketplace, more capital, and more organization than the farmers, who are simply individuals visiting the market for the day. Their lightly veiled attempt to control the market in corn brings criticism and hostility from all sides, but they manage to maintain their position despite these feelings.

The almost universal opinion among Indians and Ladinos is that corn dealers cheat in many ways: by using false-bottomed measures, by pinching the metal container when measuring (see Photo 21, p. 185), and by trying to confuse the buyer about the number of units given.[7] Many dealers also have oversize measures

---

[7] One dramatic instance of a corn dealer's trickery was reported to me by an informant who saw it during the summer of 1964. Because of increasing

for their own use in buying corn. Some farmers protect themselves by refusing all dealings with middlemen. Others, though wary, carefully measure their corn before leaving home and trade only with dealers who will accept that measurement. In this case, the farmer usually gets the going market price for his corn and leaves it to the dealer to make a profit by waiting for higher prices or cheating the consumer.

Jorge Capriata, who studied San Cristóbal corn dealers during the summer of 1965, arranged to measure a number of purchases made from dealers by regular customers buying under normal conditions. On the average, these purchases were 14.5 percent short of the official measures. Comparable purchases made from Zinacantecos selling their own corn were found to be 3.4 percent more than the official measures (Capriata 1965: 8). The excess is explained by the Zinacanteco custom of heaping corn in the official measure instead of leveling it off at the top.

Capriata found nine or ten regular dealers in the market and three or four who joined the group occasionally. He also identified a subgroup of five men who were bound by kinship and ritual kinship ties and acted together in their marketing activity. There are few data on the proportion of the market controlled by the dealers, but Capriata estimated that they sold about 90 percent of the approximately 2.5 metric tons (232 almuds) sold daily in the market during the period of his study. A large part of his study period coincided with the period of weeding, so farmers are probably underrepresented in his data. Even so, his estimate substantiates informants' opinion that dealers substantially control the sale of corn in the San Cristóbal market. The dealers buy corn in bulk from Zinacanteco producers and from other producers and middlemen

---

complaints about the dealers, the Presidente Municipal (mayor) of San Cristóbal personally inspected the market. He found a man who had bought five liters of corn from a dealer and had the purchase remeasured by a market official, who used an official one-liter measure. When the dealers were confronted with the fact that the corn measured only 4.5 liters, the oldest and most experienced of them came forward and measured the corn with the same official measure. Through four measurings, in front of the Presidente Municipal and an assemblage of onlookers, the market official repeatedly got 4.5 liters and the dealer five full measures.

in the area; but they sometimes take time out to farm themselves (see Antun's case in Appendix B), and when the price is high they may even buy from the receiving center in San Cristóbal and resell in the market.

Because of the dealers (and other factors that will be discussed in Chapter 7) the San Cristóbal market was no longer a major outlet for the surplus production of Zinacanteco farmers in 1966. However, most Apas and Nachig farmers continue to sell some corn there (see Table 5.3). Some Zinacantecos also sell substantial amounts of corn in the Chiapa de Corzo and Tuxtla Gutiérrez markets, and there are a few "private" sales between residents of Zinacantan.

Sale of Corn Futures

Many farmers who run out of cash before the harvest is in will sell a small part of their crop to someone who agrees to take future delivery. Besides family emergencies that produce a sudden need for cash, the major occasion for this sale appears to be in early August before the Fiesta of San Lorenzo, the patron saint of Zinacantan. My examination of detailed records kept by one informant who made more than twenty such purchases over two years, together with fragmentary reports of similar purchases by others, indicates that on the average a farmer who sells a future before the end of August is lucky if he receives $5.00 per almud. Given the local norms for interest, the buyer of futures is not doing exceptionally well, especially since farmers are often reluctant to deliver at harvesttime. On the whole, the sale and purchase of futures is disruptive in a small community, and one informant who bought heavily in 1966 said he would stop doing so because of the conflict it created. For all these reasons, the sale of futures, like sale to middlemen, is not a major outlet for the typical farmer's corn surplus.

Marketing Alternatives

Every farmer must choose between alternative ways of transporting and marketing his crop. For the corn that he stores to feed his family and to use as capital and seed in next year's farming, the transportation cost is determined by the location of his fields (see Tables 5.1 and 5.2). For the corn he sells, various alternatives are open to him; the volume he has available to sell and the immediacy of his needs for cash are the most important factors in determining his choice.

Table 5.8 summarizes the cash return from various marketing alternatives. There are no transport or marketing costs associated with private sale in the lowlands, and the table shows the actual cash received. For sale to the government receiving centers, transportation costs are deducted from the gross profit; and for sale in the San Cristóbal market, both transport to the highlands at harvesttime and marketing costs at the time of sale are deducted. Clearly, if a farmer based his decision on mean price alone, he would sell only to the receiving centers. The factors that make the other two outlets more attractive to some farmers are summarized below.

Private sale in the lowlands is usually to the landowner or to a middleman who buys in the fields and makes his profit either by selling in bulk to the government receiving centers or by storing the corn and waiting for peak market prices. A farmer may sell

TABLE 5.8

Marketing Alternatives

(cash received per almud minus transportation and marketing costs)

Alternative	Production zone	Mean	Range
Private sale in lowlands	all	$8.35	$7.50–9.00
Sale to government center	all	9.00	9.00
Sale in San Cristóbal market:[a]			
by Nachig farmers	2	8.25	6.85–10.65
by Nachig farmers	6	7.80	6.40–10.20
by Nachig farmers	9	8.10	6.70–10.50
by Apas farmers[b]	2	7.95	6.55–10.35

[a] The mean price received before costs are deducted is assumed to be $9.60. As explained in the text, this represents the mean for 1965–67. The ranges use the extreme monthly means for this period. Mean transport cost from the lowlands was used to construct the table.

[b] As stated in the text, Zone 6 is used by Apas farmers only for production to be sold in the lowlands, and Zone 9 is not used by Apas farmers.

all or part of his "surplus" in this way for one of three reasons: (1) he has a small surplus and does not want to invest the time and effort involved in selling to the receiving centers; (2) he needs cash to transport the bulk of his crop home or to the receiving center; (3) he is unable to deal with the bureaucracy of the receiving center and knows no one who will do it for him. In the case of small sales at relatively good prices, this alternative is convenient and offers an insignificant financial loss. Large-scale farmers who wish to sell a large part of their crop before returning to the highlands prefer to use the government receiving centers, since private sales seldom bring comparable prices.

The San Cristóbal market presents the greatest range of prices, despite the recent stabilization of prices over the annual cycle; and a farmer who anticipates that he will have to sell at the low price in this market is considerably better off selling in the lowlands.[8] On the other hand, sale at the peak price in San Cristóbal is far more attractive than the other marketing alternatives. Zinacanteco farmers keep a close watch on San Cristóbal prices during the summer months and hope to make a killing by selling when the peak is reached. However, since sales in the market are principally small-scale retail transactions, a large operator cannot hope to sell much of his surplus there even if he has the free time that such sales require.

[8] Price stabilization in recent years is discussed in Chapter 7. Although I have not studied the specific sources of corn sold in the San Cristóbal market, there is every reason to think that the corn sold there at the lowest prices is produced in the highlands or in lowland areas from which the cost of transportation to San Cristóbal is lower than it is from the area farmed by Zinacantecos.

SIX

Alternative Production Strategies, 1966

EVERY ZINACANTECO farmer must decide how much corn to seed and where to seed it. In varying degrees, both the volume seeded and the location determine how much work the farmer will have to do and how many workers he will have to hire; and these same factors are the best general predictors of what his production and profits will be.

This chapter will describe the economic implications of the various alternatives open to the farmer. My purpose here is simply to bring together the economic standards set forth in Chapters 3–5 so that the economic outcome of various farming strategies can be estimated. The strategies will be characterized in terms of location and scale of operation. For example, operations that seed two almuds and four almuds in Zone 2 will be compared with each other and with operations that seed similar amounts in Zone 6. Input variables like wages, the cost of hired workers, and transport cost will be calculated in terms of the values fixed in earlier chapters. At the outset I will also consider the effect of various uncontrollable factors on crop yields; but the more detailed descriptions will be given in terms of the normal yields set in Chapter 4.

The Farmer's Time and Capital

The farmer's expenses in corn and cash for rent, seed, hired labor, and transportation are easily calculated. But the value of his time and the cost he incurs by passing up alternative uses of the capital he invests in farming are much more difficult to estimate, because farmers spend varying amounts of time in managerial activities (e.g., recruiting workers and talking with landowners or group leaders) and because alternative uses of capital are limited. In the estimates made here, neither the farmer's work in the fields, where he could be replaced by a worker, nor his work in organization, paying rent, and similar activities in which workers are seldom if ever involved is calculated as part of the cost of production. Rather, the difference between the value of the final product and the input for rent, seed, paid labor, and transport is seen as an undifferentiated return to the farmer's labor, managerial activity, and investment of capital. The return to managerial activities and capital investment could of course be estimated by replacing the farmer's labor with that of a worker. This alternative is less appropriate than the one used, I think, because a Zinacanteco farmer normally does not seriously consider economic alternatives to corn farming. I hope that the reader who wants to make other kinds of estimates will find the data he needs in Chapters 3–5.

Selecting Representative Strategies

The nine zones into which I have divided the lowlands, combined with eight scales of operation (one to eight almuds seeded) offer the possibility of 72 farming strategies. And if we consider the different outcomes a farmer must be prepared for (good, normal, poor and disastrous yields), we are faced with 288 possible outcomes. This number will be further increased if we distinguish between transportation costs for Nachig and Apas farmers; and further complication is added by the fact that inputs for the first and succeeding years on newly cut land must be distinguished. I have chosen to reduce the complication by:

1. Making estimates for Zones 1, 2, 6, and 9 only.

2. Making estimates for 1, 2, 4, and 6 almuds seeded only.

3. Making estimates for Nachig only, and briefly discussing the differences between Nachig and Apas.

4. Reproducing only overall net return figures for the above combinations at the various yield levels.

5. Eliminating patently unsuccessful strategies that are rarely used by Zinacantecos before proceeding to the detailed description of inputs and outputs for the viable strategies.

In the remainder of this section I will try to justify the particular selections I have made.

Estimates will be made for farming operations located in Zones 1, 2, 6 and 9. Table 6.1, which shows the distribution of primary farming locations in 1966, reveals that my selection of zones to be considered is not dictated by the actual distribution of farmers. Zone 1, which is not farmed by Nachig or Apas men, is included to illustrate the economic problems of operating on land that gives very low yields. Zones 4 and 5, which show the greatest concentration of Nachig farmers, are not included in the estimates because there is so much internal variation in them. It is safe to assume that a man in Zone 2 is farming old land, and that (with the exception of one ranch) a man farming in Zone 6 is farming new land. A man in Zone 4 is probably farming old land, but there is some new land available. In my experience, those farming in Zone 5 are about evenly divided between old land and land that has recently been allowed to reforest. Thus Zones 2, 4, 5, and 6 form a continuum with Zones 2 and 6 as "pure" types.

A man in Zone 4 is typically running a Zone 2 type operation that has slightly higher yields and transportation costs than are typical in Zone 2. Men in Zone 5 may be comparable to Zone 4 operators or Zone 6 operators, except for the slight difference in transportation costs. Since land types are not systematically distributed in Zones 4 and 5—that is, two different types may exist

TABLE 6.1
Primary Farming Locations, 1966
(*from survey*)

Zone	No. of Nachig farmers	No. of Apas farmers	Zone	No. of Nachig farmers	No. of Apas farmers	Zone	No. of Nachig farmers	No. of Apas farmers
1	0	0	4	43	0	7	7	0
2	25	69	5	41	0	8	24	0
3	0	20	6	24	19	9	29	0

NOTE: For Nachig farmers $N = 193$; for Apas farmers $N = 108$.

TABLE 6.2
Amount Seeded in the Lowlands, 1966
(from survey)

Almuds	No. of Nachig farmers	No. of Apas farmers	Almuds	No. of Nachig farmers	No. of Apas farmers
1	32	18	5	14	4
2	48	32	6	22	8
3	29	31	7	1	3
4	33	11	8	14	1

NOTE: For Nachig farmers $N = 193$; for Apas farmers $N = 108$.

on the same ranch—it is impossible to estimate what a farmer is doing from the information on location by ranch name that is available. For the four zones used in the estimates, however, the type of land, and consequently the type of farming operation, is fairly consistent throughout the zone. For the purposes of rough calculation, Zone 4 may be classed with Zone 2 and Zone 5 with Zone 6.

Estimates of expenses and output will be made for operations that seed one, two, four, and six almuds. Table 6.2 shows the amounts seeded by Nachig and Apas farmers in 1966. The one-almud operation is included so that the reader can see what a small, part-time farmer faces. The three larger operations are taken as representative of the other strategies available. Besides being infrequent in the population studied, eight-almud operations are often cooperative arrangements between fathers and grown sons who are about to become independent producers. Thus they do not seem to be a viable alternative for the typical farmer.

Operations seeding two, four, and six almuds also permit relatively unambiguous decisions regarding the employment of workers, given the standards set in Chapter 4. Table 4.2 states that at many points in the agricultural cycle a farmer will do the work on up to two almuds by himself. For four almuds, he clearly must hire workers. A three-almud operation is less easily characterized. Some farmers will put in extra effort and try to work alone at some points; others will have the aid of young sons; and still others will hire workers and themselves do less work than feasible in terms of the demands of production. Given the empirically supported rule that a farmer will work the same number of days as his workers, Table 4.2 forces an estimate in terms of the last of these three alternatives. Such an estimate must inevitably be a poor approximation of what is happening in a large proportion of three-almud operations. The choice of the two-, four-, and six-almud operations as objects of the estimates permits use of the standards set in earlier chapters where they produce the most realistic results.

Since labor requirements for newly cleared land change significantly after the first year, separate estimates are made for the first and for succeeding years of farming in Zones 6 and 9, where such land is often available. The estimates for the first year are labeled 6A and 9A, and the estimates for succeeding years are labeled 6B and 9B. A farmer usually works newly cut land for at least three years before he regards his investment in clearing it as recovered, and weed incursions that require a second weeding are apt to hold off for about the same period. Therefore, overall estimates for Zones 6 and 9 are also included in the tables; they represent the means of figures for one "A" year in which clearing is done and two "B" years in which the advantages of relatively new land are still present in most situations.

In sum, eight "locations" (Zones 1, 2, 6, 9, 6A, 6B, 9A, and 9B), four scales of operation (1, 2, 4, and 6 almuds) and four yield levels (good, normal, poor, and disastrous) produce the 128 estimates of net return displayed in Table 6.3. I will give a detailed example of calculation for the reader who is interested in following out the manner in which the estimates were derived, and will then discuss the implications of the net return figures for Zinacanteco farmers.

Estimating Net Return: An Example

The process of estimating inputs and returns is perhaps best understood if inputs are broken down into four types: labor cost before harvest, labor cost at harvest, transport of people, and transport of corn. Total inputs are the sum of these four quantities. Value of the product is obtained by subtracting the corn given in rent and the corn used for seed from the total harvest and multi-

TABLE 6.3
Nachig Farmers' Net Return for Different Zones,
Scales of Operation, and Yields
(*in pesos*)

Yield & scale of operation	Zone							
	1	2	6	9	6A	6B	9A	9B
Normal:								
1	$428	$571	$691	$714	$691	$691	$714	$714
2	637	834	1268	1385	1111	1346	1219	1468
4	638	988	1950	2217	1519	2166	1767	2442
6	608	1154	2647	3015	1947	2997	2286	3379
Good:								
1	580	764	896	980	896	896	980	980
2	905	1193	1694	1921	1537	1772	1755	2004
4	1179	1678	2808	3211	2377	3024	2754	3440
6	1378	2183	3938	4533	3238	4288	3805	4897
Poor:								
1	276	426	478	516	478	478	516	516
2	381	592	852	966	695	930	799	1049
4	108	522	1205	1373	774	1421	923	1598
6	−200	432	1474	1766	774	1824	1038	2130
Disastrous:								
1	149	281	303	303	303	303	303	303
2	127	341	558	586	401	636	419	669
4	−294	51	563	685	132	779	238	908
6	−779	−287	565	735	−135	915	6	1099

plying the remainder by $9 per almud. The net return shown in the table is the value of the product minus total inputs; and it represents an estimate of what the farmer has for his own consumption or sale, assuming that he will hold enough back to run an essentially identical operation the following year. All the figures are in Mexican pesos.

The cost of labor includes wages and expenses for workers (see Chapter 3). The necessary labor input in man-days before harvest may be read from Table 4.2; this table also includes the number of days a farmer will contribute himself, so that the man-days of hired labor necessary for each step before harvest may be calculated. The labor input at harvest depends on the size of the crop, and standards for estimating it are set in Chapter 4.

A farming strategy that requires relatively few calculations may be taken as an example. Let us suppose that a farmer, Lucas, is working two almuds of new land in Zone 9. Before the harvest, he needs to hire workers only for cutting and burning the forest (see Table 4.2). He hires one worker for twenty days, paying $50 in wages per six-day week (total, $167), $7 for recruitment and maintenance, and $2 per day for food. In addition, Lucas spends $16 for the help his wife will need to prepare the food that he and his helper take to the lowlands. Total expenses for labor before harvest are thus $230. With a normal yield, Lucas would have to harvest 24 fanegas (288 almuds) of corn. The total labor required would be 36 man-days of which he would contribute 15 and two workers a total of 21. The expenses for harvest labor would thus be $175 for wages, $14 for recruitment and maintenance, $42 for his workers' food, and $12 for help in food preparation. Total expense: $243.

Transport of people during the agricultural cycle involves six trips for Lucas and three worker-trips, or nine trips at $19.20 each (Table 5.1) for a total of $173. Lucas makes trips for preparing the land, planting, weeding, doubling, and harvest; and a sixth trip is added for reseeding or looking at the fields in August. (Calculations for Zones 1 and 2 include a seventh trip for the farmer because two weedings are required in those zones.) Since different marketing strategies involve different transport costs for corn (Chapter 5), I have not charged farmers for transporting their marketable corn. Transport costs for other corn are estimated by assuming that a farmer will take home 100 almuds for family use plus one almud for each worker-day needed to run the same size operation the following year. If his crop is less than this amount, he is assumed to transport all of it home. The tables show figures that assume sale at a price providing $9 per almud plus transport and marketing costs for all marketable corn. A farmer's net profit will vary from these figures insofar as the sale price differs from this norm (see Chapter 5).

Lucas's case illustrates the one situation in which this procedure is not perfectly straightforward. Under the standards I have used, Lucas transports 121 almuds to his house at a cost of $1.10 per almud (100 for family use and 21 to pay workers for next year's harvest). Thus his expenses for transport of corn are $133. But this estimate is made for corn needed to pay workers in the coming year, when the extraordinary expenses of cutting and burning are behind him; it realistically approximates Lucas's situation, but necessarily underestimates the transportation costs he might have incurred in the previous year in storing corn to pay for cutting and burning new land.[1]

In sum, Lucas's total input for labor and transport is $230 + $243 + $173 + $133, or $779. The value of his net product is obtained by subtracting seed (2 almuds) and rent (64 almuds) from his harvest (288 almuds) and multiplying the remainder (222 almuds) by $9 per almud. This figure ($1998) minus the input for labor and transport ($779) gives the net return to Lucas's labor, managerial activities and capital investment, ($1219) shown in Table 6.3.

Eliminating Unsuccessful Strategies

The overall picture presented in Table 6.3 clearly shows that some strategies are never really viable for the Zinacanteco who wishes to support his family by farming. Success, or viability, may be defined as the production of a minimal family income of roughly $900 under normal conditions—in other words, the 100 almuds needed for family use, figured at $9 per almud.

Looking first at scale, it is clear that the operations seeding one almud never approach $900 under normal yield conditions, even

---

[1] The production process in Zones 6 and 9 could be regarded as a cycle of one "A" year followed by two "B" years. This would mean that transport costs for corn would anticipate "B" years two-thirds of the time and "A" years one-third of the time. Rather than build this complication into my estimate, I have chosen to anticipate "B" years in all cases. The difference in cost estimated is not large. In Lucas's case, the simple procedure yields costs of $133 and the more complicated procedure comparable costs of $155. The $22 difference is roughly 2.5 percent of Lucas's total input in an "A" year and roughly one percent of his total input over a three-year period.

in the best zones (6 and 9). Table 6.2 shows that 15–20 percent of Nachig and Apas farmers seed only one almud in the lowlands. But these small operations do not occupy a farmer full-time during the periods he could work on his fields, and he cannot make even a minimal living from his farming. In most cases, Zinacantecos who plant only one almud combine highland farming, wage labor, or commercial activities with their lowland farming, and the analysis made here is not capable of handling the factors that determine the relative success or failure of this mixed economic activity. Therefore, operations of one almud will be eliminated from the detailed analysis.

Looking now at location, it is clear from Table 6.3 that all operations in Zone 1 present a similar problem. Except when the crop is better than average they do not produce even minimal net returns. And under normal yields the return to scale is negative: that is, the hired labor necessary to increase scale does not pay for itself in a normal year. Since the good yield level, by definition, cannot be expected to hold over the long term, any operation of the size necessary to produce $900 net return in a good year would be uneconomic over the long run. In fact, as Table 6.1 shows, no Nachig or Apas farmers work in Zone 1. The farmers who supplied the data on which the standards for Zone 1 were based were all short-term occupants of the area (see Marian in Appendix B, for example). Despite its easy accessibility, farmers avoid Zone 1, for it is nearly impossible to make a living farming there. Thus, Zone 1 will not be considered further.

Table 6.3 displays estimates for all four of the yield levels set in Chapter 4. Although a farmer may reasonably expect normal yields, he must be ready to accept and deal with lower yields. And disastrous yields produce a negative return to scale even in Zone 9. In what follows, only estimates based on normal yields will be given; I will, however, consider some of the problems that a farmer comes up against in dealing with the possibility of lower than normal yields.

Nine strategies that are successful under conditions of normal yield will be considered in the discussion below: operations in Zones 2, 6, and 9 with scales of operation of 2, 4, and 6 almuds.

TABLE 6.4
Nachig Farmers' Investment and Income with Normal Yields
(*in pesos*)

Scale of operation and inputs	Zone						
	2	6	9	6A	6B	9A	9B
Two almuds seeded:							
Labor cost							
(pre-harvest)	$221	$77	$77	$230	$—	$230	$—
(at harvest)	100	168	243	168	168	243	243
Transport cost							
(people)	—	37	160	40	35	173	154
(corn)	123	161	133	161	161	133	133
Total inputs	444	442	613	599	364	779	530
Value of product	1278	1710	1998	1710	1710	1998	1998
NET RETURN	834	1268	1385	1111	1346	1219	1468
Four almuds seeded:							
Labor cost							
(pre-harvest)	952	571	571	996	359	996	359
(at harvest)	413	574	716	574	574	716	716
Transport cost							
(people)	—	73	282	80	70	307	269
(corn)	203	251	210	251	251	210	210
Total inputs	1568	1470	1779	1901	1254	2229	1554
Value of product	2556	3420	3996	3420	3420	3996	3996
NET RETURN	988	1950	2217	1519	2166	1767	2442
Six almuds seeded:							
Labor cost							
(pre-harvest)	1680	1074	1074	1764	729	1764	729
(at harvest)	716	964	1195	964	964	1195	1195
Transport cost							
(people)	—	105	422	115	100	461	403
(corn)	284	340	288	340	340	288	288
Total inputs	2680	2483	2979	3183	2133	3708	2615
Value of product	3834	5130	5994	5130	5130	5994	5994
NET RETURN	1154	2647	3015	1947	2997	2286	3379

Since the estimates for Zones 6 and 9 are calculated directly from the estimates for "A" and "B" years in these zones, Table 6.4 also shows details of the twelve strategies produced by combining 6A, 6B, 9A, and 9B with the three scales of operation.

Choice of Scale of Operation

A glance at Table 6.4 shows that farmers with larger operations make more profit, and that farmers in the more distant zones make more profit. I will discuss the first point here and the second in the section that follows.

It is possible to use the estimates in Tables 6.3 and 6.4 to calculate the costs and potential profits of increasing scale of operation, but I have no data on actual changes of scale made by Zincantecos in the course of farming careers. As Table 6.3 shows, the return to increased scale is negative in Zone 2, and perilously close to negative in Zones 6 and 9 when the crop is poor (25 percent below normal). Investment in a larger operation, which changes the balance of inputs away from a farmer's own labor and toward greater expenditures for hired labor and transport, is definitely a risk. On the other hand, farmers who do run larger operations are rewarded with larger incomes and consequent economic and social privileges.

Choice of amount seeded is also influenced by the capital a farmer has available, by his managerial ability and contacts, and, in large part I think, by something that must be labeled self-image. It is clear that farmers who are barely making a living from a two-almud operation have little capital available for expansion; and many farmers choose to maintain a given size of operation even after a good year that would permit them to expand. Previous failure with a large operation or aversion to the tasks of organization that face a large operator are presumably their reasons for finding a small operation satisfactory.

Choice of Location

In the context of this study, the choice of a farming location is the most important choice a Zinacanteco makes. Farmers in the distant zones have substantially higher incomes, and we are faced with the question of why all farmers do not work in these zones. If there were no differences, we would have no question thrust upon us by the idea that men seek, in their economic activities at least, to maximize income. And if the differences were small, we might ascribe a given farmer's choice to personal preferences or to membership in a social group that for historical reasons has preferred one zone to another—that is, to the noneconomic considerations that influence all economic actors. But the differences are substantial: a farmer who seeds two almuds may expect an income of $834 in Zone 2 and one of $1385 in Zone 9; and six almuds will produce $1154 in Zone 2 and $3015 in Zone 9.

If Zinacantecos are in fact "economic men," it would seem that they should all be farming in the distant zones. There are, however, three kinds of reasons why an economic man in Zinacantan might not have been working in the distant zones in 1966, when this study was made: First, the tables do not present an accurate picture of the economic situation for all farmers. Second, the cost of switching operations from the nearer to the more distant zones may present a barrier to change. And finally, the opportunities available in the distant zones are relatively new ones, and Zinacanteco farmers face the problems common to such periods of change. Each of these reasons explains part of the disparity between the profits shown in Table 6.4 and the distribution of farmers shown in Table 6.1. I will discuss them in turn.

On the whole, Nachig farmers do not farm in Zone 2. Rather, they are heavily concentrated in Zones 4 and 5, where crops are apt to be somewhat better; and even though they face a similar labor investment and higher transport costs than in Zone 2, they are likely to do better than the outcomes shown in Table 6.4 for Zone 2. Thus part of the answer for Nachig farmers is that the real differences are not as great as those shown by the extremes displayed in the table. For Apas farmers, Zone 2 itself is actually a bit better than the results shown in the table, for their transport costs are lower than those of Nachig farmers. Even so, the difference is barely $20 for a man seeding two almuds. The fact that Apas is not on a road is more important, for this makes the cost of travel to a more distant zone exceedingly high. Still another part of the answer, especially for the small operator, may be that the labor-intensive Zone 2 will suit his personal situation if he has "free" labor available through either extraordinary personal effort or the help of young sons. The more distant zones are capital-intensive in the sense that there is no substitute for the cash that must be paid for travel.

The initial costs of switching from the nearer to the more distant zones may also influence the distribution of farmers. The new fields require high investment in wage labor for clearing the land and in transportation, and the bulk of these investments are required early in the year. Hence a small-scale farmer who retails his own corn must sell in the worst period of the year in order to meet the cost of opening a distant field. A Nachig man farming two almuds in Zone 2 must invest $444 to maintain his Zone 2 operation the next year; but the $779 required for the first year of farming two almuds in Zone 9 may be more than he can afford, even if the long-run benefits are clear to him.

These and other economic reasons, separately and in combination, help to explain why all farmers are not working in the distant zones. But another large component of the answer is found in the history of the changing situation faced by Zinacanteco farmers and in the very nature of change situations. That is, the choice of locations cannot be completely explained in terms of the synchronic approach that has so far been the focus of this study. In general terms, the historical component of the explanation is simple: the conditions described for 1966 and used to make the estimates displayed in Table 6.4 did not exist a few years ago; farming in Zone 9 did not offer a great economic advantage until recently, and Zinacantecos have been moving to more distant zones as advantages have developed.

It is easy to assert the importance of the fact that things are changing, and most of us are probably disposed by personal and scientific experience to let the question rest and look at the details of the change process. But the economic model taken in its barest form does raise a question. Why did the change not take place instantaneously—in 1964, say, when the advantages of Zone 9 became available to Zinacantecos? In the broad context of experience with change this question cannot be taken seriously, for we know that this kind of change does not take place instantaneously. But it does raise the issue of the manner in which the increments that make up the process of change are themselves produced.

The diachronic description and analysis in the chapters that follow focus on this issue. Chapter 7 concentrates on the aggregate response of Zinacanteco farmers and shows movement toward more distant zones during the decade ending in 1966. Although this process may be seen as a continuous flow toward the opportunities offered in the more distant zones, it is nevertheless made up of individual decisions to move or not to move; and the individual is subject to the various uncertainties about economic outcomes that I suggested in Chapter 1. The outsider's view of Zinacanteco farming that is summarized in Table 6.4 does not distinguish between the degree of certainty of the various profit levels displayed. But a Zinacanteco does not have the information shown there, and he cannot afford the luxury of ignoring the relative newness and scantiness of the evidence for the economic superiority of the more distant zones. If he chose to move, though uncertain about economic outcomes, he had to do so on the basis of noneconomic considerations that made it desirable to take the chance. In Chapter 8, I will try to show why men at different places in the economic hierarchy might have found it more or less desirable to move to distant zones even though the economic outcome of the choice was uncertain. Though it is conceptualized in terms of a general maximization framework, their behavior is alien to the economic man, who can make choices only when he knows a great deal about economic outcomes.

SEVEN

Economic Change, 1957–1966

CHANGE HAS dominated the world of Zinacanteco corn farmers for more than ten years. They have been farming in new areas and selling in new markets in response to opportunities for better yields and better prices. Ten years ago no Apas farmers worked fields farther from their hamlet than Zone 3; now almost 20 percent of them farm in Zone 6. In the decade 1957–66 the proportion of Nachig farmers working fields more distant than Zone 6 increased from less than 5 percent to more than 30 percent. Ten years ago there were no government receiving centers; today, more than 40 percent of the farmers studied sell some of their crop to the government. The differences between the traditional patterns of farming and marketing and those typical of farmers working distant fields and selling to receiving centers represent profound changes in both the economics of farming and the social relations implicit in the work. These changes were generated by factors beyond the control of Zinacantecos: the process began with the national government programs that provided new roads and new markets; these programs, in interaction with the local situation, produced a new level of prices; and Zinacantecos responded by changing their farming and marketing practices. This chapter will discuss the changes in detail.

Most observers would also find continuities in the situation of the Zinacanteco corn farmer, and these should not be minimized simply because the focus of this part of the study is change. First of all, Zinacantecos are still farming corn; and few of them have shifted to other occupations. Moreover, the organization of work groups, the relations with landowners, the tools used, and the relations between employer and worker are very much as they were in 1957. This fundamental stability allows my description of change over a ten-year period to proceed in terms of the synchronic picture given in the chapters above. All the elements of the traditional pattern still exist in current Zinacanteco practice as it is described in Chapters 2 to 5; but with each passing year fewer and fewer Zinacantecos adhere to them.

For the most part, the data supporting the broad outline of change I have given here are in the form of maps and tables. In what follows, the text is more often supplemental to the maps and tables than vice versa.

New Roads: Changes in Transportation

The maps on pp. xii–xiii outline the gradual development of the road system. They show how areas that were inaccessible in 1957 quickly became areas that Zinacantecos could reach by truck, and how nearby areas that required day-long walks and mule transport of crops in 1957 now offer the farmer a choice of transport. The maps cannot express, however, the subtle gradation from footpath to superhighway. As every student of agriculture in developing countries knows, it is not always easy to tell just when a road is a road. A road that is passable by sedan for only two months each year may be passable by pickup truck for six months and by jeep or large truck for eight or nine months. A road that one driver considers passable may be considered too likely to damage the same vehicle by another driver. On my maps, the all-weather roads are ones on which dependable transportation is available during the entire year. The dry-season roads are ones over which trucks can travel at harvesttime. During the summer wet season, farmers who go from Zinacantan to the lowlands by truck often walk the last few miles to their fields.

There are four important roads. (1) The paved Pan-American Highway runs from Mexico City and Tehuantepec through the state capital of Tuxtla Gutiérrez, up the mountains to San Cristóbal de Las Casas and down the mountains again to the city of Comitán and the Mexico-Guatemala border. (2) An all-weather road leaves the Pan-American Highway near Chiapa de Corzo and runs along the river to Acala, Flores Magón, and Venustiano Carranza. (3) Another all-weather road leaves the highway at Amatenango del Valle and goes through Villa Las Rosas and Soyatitán to Pujiltic. (4) Finally, an all-weather road leaves the highway to the west of the river and goes towards La Concordia.

The Pan-American Highway first became passable between Tuxtla and San Cristóbal in the late 1940's. Before that, a very rough gravel road that is now abandoned and impassable connected Tuxtla and San Cristóbal along a route that ran through Ixtapa and Zinacantan Center (Hteklum). In those times Indians did not travel by motor vehicle. By 1957 the Pan-American Highway was a first-class paved road that ran through the highlands all the way to Comitán.

Work on the road to Venustiano Carranza through Acala began in 1959 (ANDSA 1964). A Zinacanteco farmer reported that four major bridges were missing when he brought his harvest out from a point beyond Acala in late 1961; and I found it a very good gravel road broken by a few difficult fords when I drove to a point short of Acala in the summer of 1962. By harvest season in 1962 the bridges were apparently finished. Before the road was built, the residents of Acala and the many small settlements travelled to Chiapa de Corzo and Tuxtla by foot and horseback, or in the large dugout canoes that are still used to cross the river in many places. Residents of settlements near Flores Magón speak of taking three days to get to Tuxtla before the road was built; the dry-season road that connected the area with Venustiano Carranza at that time was apparently an impractical way to get people or crops to the outside world. These same people remembered the opening of a practical route through Acala for the 1964 harvest season. As the maps show, the road through Acala to Venustiano Carranza was completed by 1966. It lacked no bridges, and at least one sedan taxi, as well as buses, ran from Tuxtla to Venustiano Carranza daily.

The main part of the road from the Pan-American Highway to

Pujiltic was finished earlier than the Acala road. In his study of Villa Las Rosas, Hill states (1964: 10): "Las Rosas became linked to [the] national transportation network in 1955, largely to provide a modern transport route for the sugar refinery at Pujiltic."[1] The all-weather road from Pujiltic to the Socoltenango area was built to reach the construction site of a large irrigation project beyond Pujiltic and has been open for about two years. The dry-season road that goes to Porvenir is an older route, but local residents say that it is only in the last two years that it has been consistently open all the way to Porvenir. At present it is impassable to trucks only during September and October. Before these last two roads were completed, the area beyond Pujiltic was largely cattle country, although local people farmed corn there.

The link from Pujiltic to Venustiano Carranza includes a major ford of the Río Blanco, and at several places in flat areas the route wanders through a maze of dirt tracks. But a bridge across the Río Blanco is planned for the near future, and with it will no doubt come other improvements that will make the lowland route from Chiapa de Corzo to Amatenango a complete all-weather road on which it will be possible to avoid the torturous mountain drive along the highway through San Cristóbal.

The route from the Pan-American Highway to La Concordia on the west side of the river is less important to Zinacantecos than the roads mentioned above. Very few Apas and Nachig men farm in the area to which it gives access, but a number of men from the hamlets at the western edge of Zinacantan do so. Work on the road began in 1959 (ANDSA 1964), and residents of the area report that it was passable to La Concordia by the early 1960's.[2]

Along with the improved roads came a sharp increase in the number of motor vehicles in the area. Table 7.1 shows the figures

[1] Local people work in the sugar refinery, around which a sort of company town has developed. Zinacantecos pass through the complex of buildings, roads, and irrigation ditches to reach the corn lands beyond.

[2] This road is mentioned only to make the ethnographic record complete. The distance to the new fields it opens up is too great on foot for Nachig people, and the fares for motor travel are much higher than those from Nachig to Zone 9. The balance of distance and cost of transport apparently shifts for Zinacantecos in the western hamlets of the township.

TABLE 7.1
Motor Vehicle Registration in Chiapas

Year	Autos	Trucks	Year	Autos	Trucks
1949	1,088	1,102	1961	3,697	4,355
1957	2,329	3,076	1962	3,666	4,086
1958	2,516	3,232	1963	4,313	4,951
1959	3,113	3,800	1964	4,992	4,956
1960	3,121	3,810	1965	5,618	6,124

for the state of Chiapas for 1949 and for the period of 1957–65; and my guess is that the rate of increase for the area of study is at least as high as that shown for the state as a whole. Though the pack animal, oxcart, and dugout canoe have not completely outlived their usefulness, there is no doubt that a calculation in terms of ton-miles or some similar standard would show that trucks have come to dominate the transport of corn. This has affected the Zinacanteco farmers in a number of indirect ways. When I first worked in Zinacantan in 1960, the assertion that a man was wealthy was often backed by a report of how many mules he owned, but this is no longer a general standard. In the past, every farmer aspired to own several mules, and no one who had less than one or two was considered properly established. A truck is still a specialized investment that no farmer considers a normal part of his equipment.[3]

The shift away from pack transport has also affected the off-season occupation of many prosperous Zinacantecos. Traditionally, men who owned a half-dozen or more animals transported their own corn and hired themselves out to transport the corn of others. In addition, they did contract hauling of crops, principally coffee, from the hinterlands to San Cristóbal. New roads in the highlands, and to some extent air transport, are rapidly decreasing the oppor-

[3] In 1960 one pickup was owned by Zinacantecos, and by 1966 a number of groups had bought trucks. But the number of trucks owned wholly or in part by individuals and groups of Zinacantecos was then nowhere near a dozen. All of the vehicles are constantly in service and are attended by men who depend on trucking rather than farming for the major part of their income.

tunities for such work. The case of Guilliermo, the leader of a group farming at Porvenir, seems typical of this change. Until 1964 he owned a horse and four mules. He transported his own corn from his various farming locations (Zones 4, 5, and 6), and says that he was continually asked to transport the crops of other Zinacantecos. He also had a permanent arrangement with a San Cristóbal man who traded in coffee futures; and in each year through 1964 he hauled coffee to San Cristóbal in the off-season (January to May). When his horse and two of his mules died of old age Guilliermo decided not to replace them, since, he says, everybody farms close to the road now. And he has moved his own fields to an area that is too far from his home to make pack animals useful.

These changes have not affected all Zinacantecos equally. There is still no motor road to Apas; and even if there were, farmers whose cornfields are just below their homes might still prefer to use pack animals. However, the establishment of government receiving centers in the lowlands has made it possible to sell profitably at harvesttime and completely avoid the transport of much corn to the highlands. Thus, even though Apas farmers live off the roads, the usefulness of their pack animals may be substantially reduced by the marketing changes that were made possible, in part, by the new roads.

New Markets: The Receiving Centers

Rapid expansion of the government corn-buying program paralleled the expansion of the road system. By 1966, receiving centers had been established within easy access of all areas farmed by Zinacantecos. For the most part, the development of the receiving centers may be read directly from the maps and from Table 7.2. The single special case is the relation of the centers in Flores Magón and Venustiano Carranza. The center in Venustiano Carranza, which purchased increasing amounts of corn from 1959 to 1962, operated in a temporary canvas warehouse that was set up each year for the buying season. When a permanent warehouse was ready in Flores Magón in 1963, the Venustiano Carranza

TABLE 7.2

Purchases by Government Receiving Centers, 1959–65 Crop Years

(metric tons of corn)

Location	Year[a]					
	1959	1961	1962	1963	1964	1965
Acala	—	—	6,549	6,837	7,357	8,168
Chiapa de Corzo	12,139	12,387	6,855	7,075	7,221	5,540
Colonia 20 de Noviembre	—	—	—	6,427	8,693	11,058
Flores Magón	—	—	—	10,559	11,535	11,514
El Brillante	—	6,252	8,428	12,651	13,499	11,672
La Concordia	—	—	—	—	—	7,518
San Cristóbal	2,592	1,153	1,235	4,410	4,996	6,088
Venustiano Carranza	2,563	3,388	7,447	—	—	4,899
TOTAL	17,294	23,180	30,514	47,959	53,301	66,457

NOTE: I am indebted to Lic. Bernardo Serra Altamira, chief of economic studies for Almacenes Nacionales de Depósito, S.A., and his staff for their help in compiling these figures from unpublished reports. No data for 1960 were available. Data for 1966, which were received much later by mail, show smaller total purchases than 1965. I cannot be sure that they were tabulated in a manner parallel to those shown above and therefore have not included them in the table.

[a] From 1961 on, the receiving centers typically began buying on November 1 and closed no earlier than the middle of April in the following year. The dates on the table represent opening dates—i.e., 1962 indicates the 1962–63 buying season for corn seeded in 1962.

center was discontinued; it reopened in 1965, after construction of a permanent warehouse had begun. The opening of new receiving centers is reflected in the decline of purchases at nearby centers. For example, purchases at the Chiapa de Corzo center fell off sharply when the Acala center was established in 1962; and in 1965 purchases at the El Brillante and Flores Magón centers fell off slightly when the Concordia and Venustiano Carranza centers went into operation.

The sharpest increase in the quantities sold to the receiving centers occurred between the 1962 and 1963 crop years. This rise coincided with the opening of new receiving centers; but it was chiefly the result of a dramatic increase in the government's buying price in fall 1963. The increase, from $800 to $940 per metric ton (before discounts), was explicitly intended to expand purchases and improve the relative position of the agricultural sector in the Mexican economy.

It is difficult to estimate the percentage of total corn production purchased by the government in the area of study, for figures on production broken down by township were not available. On the whole, Chiapas produces a substantial surplus over its own needs, and the centers there are expected to ship most of their purchases to corn-poor states. For 1960, the records of the Plan Chiapas (Plan Chiapas 1962), a government program to increase corn production in the state, shows a production of 273,000 tons, a local consumption of 168,226 tons, and an export (by both the government and the private dealers) of 104,774 tons. For the 1963 crop year, the Chiapas branch of the department of agriculture estimated the total corn harvests at 420,000 tons and purchases by receiving centers at 150,000 tons (*El Universal* 1964). In the same report, the director of the branch estimated that the government purchased 30,000 of 260,000 tons produced annually in the late 1950's, before the Plan Chiapas went into effect. Even if local production has risen faster than local consumption because of the plan, the purchases made by the receiving centers in the middle 1960's were probably more than half the surplus corn on the local market; thus the centers provide an effective price floor at harvesttime, when many small farmers need to sell a substantial part of their crop.

Zinacantecos, although they concentrate on corn farming, do not produce enough to influence the local market substantially. A very rough estimate of their total production in the lowlands might be 5,000 tons, and they produce considerably less than that in the highlands.

The Price of Corn

Because of the receiving centers and the roads that made them practical, the price paid Zinacantecos for their corn has risen and stabilized dramatically during the period under study. Before the establishment of the centers, the major outlets for the corn grown by Nachig and Apas farmers were two: retail sale in the San Cristóbal marketplace,[4] and bulk sale in the lowlands to middlemen

---

[4] Sale to dealers in the San Cristóbal market is a recent thing. Before 1960 there were fewer dealers, and they did not have the dominant role described in Chapter 5.

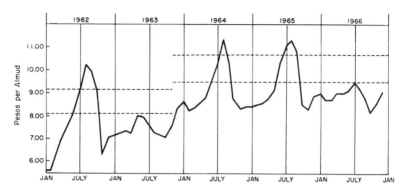

FIGURE 7.1. Corn Prices in San Cristóbal

and landlords acting as middlemen. When the centers appeared prices in the two older outlets responded to the prices offered by the government.

Figure 7.1 depicts the mean price of corn in San Cristóbal for each month from 1962 through 1966. The curve is based on daily observation of the price arrived at after the early morning fluctuation that is sometimes characteristic of the market.[5] When the effective price paid by the receiving centers went from $8.15 to $9.50 per almud in late 1963, the price in San Cristóbal quickly followed. Table 7.3, which summarizes Figure 7.1, shows this radical change even more clearly than does the figure. Obviously, not every Zinacanteco who sold in the San Cristóbal marketplace during this period realistically considered the receiving centers as an alternative outlet for his product; nevertheless, the centers' effect on the market for corn reached everywhere in the system because of the middlemen who were prepared to buy corn in small lots and resell it to the centers.

[5] An assistant of mine, Teódulo Martinez Osuna, recorded corn prices daily beginning in 1962. Though he often left San Cristóbal for a few days at a time because of his full-time job, his family remained in his house near the market; and his wife, who makes tortillas for sale in the market, continued to buy corn daily. His records were occasionally suspect, but I am confident that they are very accurate on the whole. Monthly means were calculated by taking the mean of the high and the low price reported for each week (with odd days counted in the preceding month).

TABLE 7.3
Corn Prices in the San Cristóbal Market, 1962–66
(*pesos per almud*)

Category	Year				
	1962	1963	1964	1965	1966
Mean (annual)	$7.70	$7.55	$9.15	$9.40	$8.90
High month	10.30	8.10	11.30	11.35	9.50
Low month	5.55	7.10	8.25	8.30	8.20
High/low ratio	1.83	1.14	1.37	1.37	1.16

The prices in the San Cristóbal marketplace and the prices in the receiving centers do not match exactly, and San Cristóbal prices often go below the centers' buying price of $9.50 per almud or above their selling price of $10.70 per almud. The fact that the center is not effectively open to all prospective sellers may be seen as responsible for sales below its buying price. Middlemen who are willing to invest in corn and go through the procedures necessary to get it into the receiving centers are able to take a profit for doing so when corn is plentiful.

In 1964 and 1965 when the peak price in the marketplace went substantially above the centers' selling price, the San Cristóbal center advertised its selling program. Many consumers bought from the center, but most of them could not or would not raise the cash to make the minimum purchase—a bag that is roughly 25 times a typical family's daily purchase in the marketplace. After repeated efforts to control the retail price through legislation had failed, the municipal authorities of San Cristóbal decided to set up their own retail outlets to resell corn from the centers whenever market prices were high. These municipal outlets, which operated on several occasions in recent years, no doubt helped many families; but they did not seem to change the market price substantially. Many consumers have personal ties to retailers that cannot be broken without incurring social costs, and others are unwilling to forgo the credit opportunities that small retailers offer to favored customers.

The overall stabilization of prices resulting from the creation of the receiving centers is also important to many Zinacanteco farmers. The ratios of high to low monthly prices in Table 7.3 show this stabilization after 1962. Informants say that the prices prevailing in 1962 and the fluctuation in that year are typical of the market's behavior for a number of years before 1963. In years of extreme shortage, prices soared even higher for short periods; and occasionally, at harvest time, corn was sold for as little as $4.80 per almud in the San Cristóbal marketplace. On the whole, however, the range of prices shown for 1962 encompasses the fluctuation usually found in the traditional market. Unfortunately, I have no systematic data on the San Cristóbal market for earlier years; but I lived in the area almost continually from September 1960 to January 1962, and my own experiences support informants' statements that 1962 was not an unusual "traditional" year. Given the data I have, the fluctuation of the high to low price ratio between 1.14:1 and 1.37:1 during the years 1963–66 is unexplainable in anything but ad hoc terms. However, the overall trend toward higher and more stable prices in San Cristóbal is clear.

Lowland prices have also risen sharply during the period under study. Most Zinacantecos who sell to private buyers in the lowlands do so immediately after harvest, when the annual price cycle hits its lowest point, so price stabilization is especially important to them. Since systematic data on lowland prices before 1963 are too scanty to provide substantial support for these claims, I will confine my argument for them to presenting the report of one informant from a dramatically affected area. Other bits and pieces of information that I have are consistent with this report.

My informant is a resident of Flores Magón who owns land in Zone 6 and rents to two large groups of Zinacantecos, one from Apas and one from Nachig. When he began renting to the Nachig group in 1960, he sometimes bought corn from the farmers who worked his land, buying at a bit less than $3.00 per almud and selling to truckers at $3.50 to $4.15. These truckers braved a very poor dry-season road from Venustiano Carranza (not shown on maps) and took the corn to major markets as distant as the state of Oaxaca. My informant reported that this trade was an irregular thing, and that he had usually had a hard time disposing of 50 fanegas (600 almuds) to incoming truckers. When the receiving center was established in Acala for the 1962 buying season, there was still no road out in that direction, but the price of corn immediately climbed to $5.00 and more per almud. And as soon as a receiving center opened in Flores Magón, during the 1963 buying season, the price paid by my informant and other local middlemen went immediately to more than $8.00 per almud ($100 per fanega).

In sum, the prices now paid by middlemen in the lowlands are much closer to prices paid elsewhere than they were a few years ago. Table 7.7 (in the section on marketing) shows a predictable trend toward sale to the receiving centers for larger amounts and sale to middlemen for smaller amounts. A decade ago, only farmers who were desperate for cash sold to middlemen in the lowlands at harvest time, and any successful farmer had to transport the bulk of his crop to his home. Today, lowland sale is a viable economic alternative for almost any farmer.

New Fields and New Markets: The Response to Changing Opportunities

The aggregate response of Zinacanteco farmers to the changes described above has been dramatic. As opportunities have appeared, they have adopted new farming locations and the new markets. Table 7.4 tells the story of movement to new fields better than it can be told in words. Over the last three or four years especially, there has been a rapid shift from farming in the nearer zones to farming in zones where the farmer must depend on motor transport to reach his fields and government receiving centers to most economically market his crop.

The importance of roads is reflected in the differing responses of Nachig and Apas farmers. Nachig farmers, who live near the Pan-American Highway, moved to the more distant locations in greater numbers; and they did it sooner than Apas farmers, whose homes are an hour away from the nearest road. The initial movement of Apas farmers out of the traditional zones (1964) was confined to a single work group under the leadership of one man, and even today only two Apas groups farm in Zone 6. By contrast,

TABLE 7.4
Primary Farming Locations, 1957–66
(from survey)

| | | Nachig | | | | | | | | | Apas | | | |
| | | | | Zone | | | | | | | | | Zone | |
Year	N	2	4	5	6	7	8	9	Year	N	2	3	6
1966	193	13%	22%	21%	12%	4%	12%	15%	1966	108	64%	19%	18%
1965	181	13	26	23	14	5	10	9	1965	110	66	17	16
1964	169	17	29	21	15	2	8	8	1964	109	73	12	15
1963	166	23	28	19	15	2	6	6	1963	107	93	7	—
1962	153	26	29	19	17	4	1	4	1962	101	97	3	—
1961	140	26	30	19	16	6	—	4	1961	96	96	4	—
1960	139	34	26	19	16	5	—	1	1960	85	96	4	—
1959	137	42	25	18	10	4	—	1	1959	80	89	11	—
1958	131	46	28	15	9	1	—	1	1958	76	84	16	—
1957	124	44	28	15	10	1	—	1	1957	69	82	17	—

NOTE: Of the 1,533 farmer-years shown for Nachig 33, or about 2.2%, were not actually worked in the location indicated. Ten cases (spread over eight years) in Zone 1 were classified with Zone 2. Seven cases (spread over seven years) in Zone 3 were classified with Zone 6 because both Zones 3 and 6 are at the limit of the traditional transportation system. Sixteen cases (spread over nine years) involving motor transport to locations west of the Grijalva River were classed with Zone 7 (3) or Zone 9 (13) depending on the distance. The effect on the table is negligible except for those that fall in Zone 9 between 1957 and 1960. In fact, no farmer in the sample farmed in Zone 9 during those years. Of the 941 farmer-years shown for Apas, one was worked in Zone 1 and classified as Zone 2.

the spread of Nachig farmers has involved a much larger and more diverse part of Nachig's population.

The primary locations presented in Table 7.4 are always the more distant fields of farmers who work on two separate ranches. Table 7.5 shows that the percentage of farmers working two locations has increased substantially over the last few years. And, because of the difficulties in transport encountered by Apas farmers who work outside the traditional zones, a much higher proportion of them have other fields near their homes. The complete data show that 70 percent of the Apas farmers who have used Zone 6 have also maintained fields in Zone 2, from which transport to their homes is relatively economical. By contrast, only 26 percent of the Nachig farmers in Zones 7–9 have maintained secondary fields nearer home.

The increasing sale to government receiving centers in the years

TABLE 7.5
Secondary Farming Locations, 1957–66
(*from survey*)

	Nachig farmers		Apas farmers	
Year	With two locations	2d location nearer home	With two locations	2d location nearer home
1966	20%	16%	25%	21%
1965	15	12	30	21
1964	12	9	28	16
1963	12	8	12	<5
1962	10	7	<10	<5
1961	<10	5	<10	<5
1957–60	<10	<5	<10	<5

since they have been established is shown in Table 7.6. Here, too, Apas farmers have not changed their practice as much as those from Nachig. This may reflect a conservative tendency in the Apas population; but I think that the differences between the hamlets are more readily explained by the fact that the advantages of the receiving centers have become more important to Nachig farmers, who are more often working farther from home. Table 7.6 gives no indication of the volume of corn sold at each of the outlets, although it shows that the proportion of farmers using each of the traditional outlets has not decreased. Table 7.7, which does include information on volume sold, clearly documents the predictable trends in lowland sales of corn. Larger sales go to the receiving centers, and smaller sales to the private dealers and landlords. Many farmers who sell the major part of their crop to the receiving centers nevertheless obtain ready cash by selling small amounts to private buyers who are willing to accept delivery in or near the fields.

On the whole, the response of Zinacantecos to the new opportunities is clear. And, of course, this response has continued since my survey was made. Farmers have continued to move to the new locations as the new opportunities become ever more apparent. Although I have used words like "rapid" and "dramatic" in describing the Zinacantecos' response to the new situations, it should be

TABLE 7.6
Sales Outlets Used by Nachig and Apas Farmers, 1962–66
(*from survey*)

| | Outlet | | | | |
Year	San Cristóbal market, to consumer	San Cristóbal market, to dealer	Government receiving center	Private sale in the lowlands	Speculator, before harvest
Nachig:					
1966 ($N = 191$)	88%	12%	49%	65%	24%
1965 ($N = 179$)	87	12	47	61	16
1964 ($N = 168$)	87	8	41	57	18
1963 ($N = 165$)	89	10	32	57	13
1962 ($N = 152$)	88	13	11	62	13
Apas:					
1966 ($N = 108$)	55	16	33	58	51
1965 ($N = 110$)	62	14	37	55	49
1964 ($N = 109$)	68	11	34	54	42
1963 ($N = 107$)	68	15	22	57	37
1962 ($N = 101$)	66	14	12	49	46

NOTE: Percentages are percent of all farmers surveyed.

clear that these terms represent a completely unsupported implied comparison. Some observers might want to claim that the response was in fact sluggish, or that Zinacanteco farmers have been irrational in resisting the proffered opportunities. I do not think any fieldworker who has been close to the mood of Zinacantecos as they discussed the complications and uncertainties of the changes they have made would characterize the situation in this way; but my data do not permit any statement about the rate of change of the Zinacanteco aggregate in comparison with other aggregates. Here, it can only be concluded that the response to the opportunities offered by new roads and new markets has been clear and substantial, and that it is continuing.

Knowledge of Possibilities

In any change situation, it is possible that individuals who do not adopt new behavior patterns are restrained by ignorance; that is, they may not realize that desirable opportunities exist. In fact,

TABLE 7.7

Corn Sales in the Lowlands, 1963–66

(*from work histories*)

	Means		Ranges	
Date and buyer	Price per fanega	Volume in fanegas	Price per fanega	Volume in fanegas
1963:				
Private ($N = 3$)	$102	8	$100–105	2–11
Receiving centers ($N = 3$)	114	6	114	3–8
1964:				
Private ($N = 3$)	90	5	80–100	3–6
Receiving centers ($N = 2$)	114	8	114	5–11
1965:				
Private ($N = 6$)	96	7	90–110	3–11
Receiving centers ($N = 4$)	114	13	114	6–23
1966:				
Private ($N = 6$)	100	3	90–108	2–4
Receiving centers ($N = 4$)	114	30	114	23–45

NOTE: Price given per fanega (12 almuds) to simplify the numbers; $114 per fanega = $9.50 per almud.

the many studies of the diffusion of innovations that are framed in terms of knowledge (see Rogers 1962) proceed on the assumption that knowledge of an opportunity is crucial to its acceptance. Although knowledge is important, complete knowledge based on experience is by definition impossible in a situation of newness. Chapter 8 will stress the view that action under uncertainty is a crucial characteristic of such situations of change.

Nevertheless, it is important to have some idea of how widely diffused the knowledge of developments in new areas is among the Zinacanteco farmers studied. Zinacantecos often discuss the quality of land, transport costs, and relations with landowners that apply to specific ranches in areas where they have not worked. And some farmers visit many parts of the lowlands in search of fields that may be better than those they are working. This kind of information and experience is important to every farmer, and I have never met a Zinacanteco who did not at least feign interest in any new details I might have about unfamiliar lands. Since any reasonably efficient attempt to measure the extent of such knowledge in the entire study population had to involve questions about universally recognized new features of the farming situation, I selected the receiving centers as the objects of my questions.

Table 7.8 (which should be studied in conjunction with the maps) shows the results of those questions on the survey. Respondents were asked if there were a receiving center in each of seven well-known lowland locations; and if they answered yes for a given location, they were asked whether they had been there and seen the receiving center. As a check on the care with which questions were being answered, I included two locations that do not have receiving centers in the list. The responses show that almost all farmers know the locations of receiving centers and report having seen the ones in nearby locations. Although more than 10 percent said there was a receiving center in the distant town of Soyatitán, when there was not, very few responded positively (incorrectly) to the questions about nearby Chiapilla. The table also reflects the

TABLE 7.8
Farmers' Knowledge of Receiving Centers
(from survey)

Location	Nachig responses[a]				Apas responses[b]			
	Had seen center	Knew of center, had not seen it	Did not know	Certain there was no center	Had seen center	Knew of center, had not seen it	Did not know	Certain there was no center
Chiapa	94%	4%	2%	0%	91%	5%	4%	0%
Acala	93	5	2	1	99	1	0	0
Colonia 20	45	17	35	4	73	21	4	2
F. Magón	86	9	5	0	71	27	1	0
Venustiano C.	74	13	10	3	22	32	35	11
Chiapilla	3	0	2	95	0	0	10	90
Soyatitán	12	1	41	46	7	7	80	7

[a] $N = 191$ except for Venustiano Carranza, where $N = 190$. Responses were not recorded for two (and in the case of Venustiano Carranza, three) of the 193 farmers who had lowland fields in 1966.
[b] $N = 103$ except for Colonia de 20 Noviembre, where $N = 100$, and Flores Magón, where $N = 102$. Responses were not recorded for five informants at all, and three more of the 108 farmers who had lowland fields in 1966 gave incomplete responses.

different geographical orientations of Apas and Nachig farmers, the Apas farmers being uncertain about Venustiano Carranza in Zone 8 and the Nachig farmers being uncertain about Colonia 20 de Noviembre in Zone 3. On the whole, it may be concluded that simple ignorance of opportunity is not a major reason for the less than complete shift to the new farming areas.

False Starts and Fiascos

The patterns of change described above are so much the predictable ones, and the description is so abstract and gushingly positive, that the process of change in Zinacantan may appear to have been smooth, and free of the false starts and fiascos that are typical of change situations. This is not the case, though it is impossible to make any comparable assessment of how smooth or difficult the changes have been. In the next chapter, where the individual and not the aggregate is the unit of analysis, the reader may get a more realistic sense of the ambiguities involved in the changing situation faced by Zinacanteco farmers. Here, I will very briefly review their experience with two technological innovations that have not gained wide acceptance.

Chemical weed killers (of the 2-4-D type) have been actively merchandised in the state of Chiapas for several years, and some Zinacantecos used them as early as 1963 or 1964 (Cancian 1965b). These chemicals kill broadleaf weeds, leaving corn and other grasses unaffected; and where the weed competition is principally broadleaf, they can greatly reduce a farmer's labor investment.[6] Although I talked with only a few Zinacantecos who have used weed killers, it is clear that the practice can produce substantial savings under the right conditions; and the use of weed killers eliminates the immense labor recruitment problem that many farmers face during the June and July peak work periods.

When land is newly reclaimed from forest there are few weeds, and most of these are broadleaf; but after years of use, or when the land is cropped before it is completely reforested, much of the

---

[6] Weeding with hoes also cultivates a field, whereas weed killers leave the soil undisturbed. But most agronomists I asked considered cultivation far less important than the elimination of weed competition.

weed competition comes from grasses. These circumstances produced some classic fiascos when the weed killers were first introduced. Some farmers thought the chemicals would dispose of all
weeds and applied them to old fields where the principal compe-
tion came from grasses. By the time they realized their error, it was
too late to control the problem by conventional weeding. My impression is that by 1966 the tales based on these incidents had wide
enough circulation so that farmers who were considering the use
of weed killers were aware of the conditions under which they
might be used. Although more farmers could profitably use chemical weed killers, most Zinacantecos work land for which they are
not appropriate or economic.

Hybrid corn seed, which is distributed by government agencies
rather than the commercial houses that sell chemical weed killers,
is also used by some Zinacanteco farmers.[7] Hybrids like Rockefeller 507 must be purchased anew at least every few years, and
some fiascos occurred when farmers insisted on taking seed corn
from their crops. But in general the farmers' stories about hybrids
show a more deliberate approach to the advertised virtues of the
new seed. In at least one case, an Apas farmer tested hybrid and
local seeds on similar plots and proved to his satisfaction that yields
from hybrid were not significantly greater than those from the
seed he had selected from his previous year's crop. The marginal
land farmed by Zinacantecos often does not contain the substantial
nutrients necessary for optimal production from hybrids; and fer-
tilizers, which have been experimentally introduced in some highland communities, are apparently impractical on the typically
sloping Zinacanteco fields.[8]

Moves to the new farming area have also produced surprises for

_____

[7] Zinacantecos use Rocamex, a "stabilized hybrid," and (according to informants) a few others like Rockefeller 502 and 507. My work histories include 88 farmer-years with information on seed type. Of these, 16 involved
some hybrid seed; and in 12 hybrid seed was used exclusively. Since the cases
for which I have no information are more likely to involve native seed types,
the rate of hybrid use is probably not as high as these figures imply. Farmers
take up and drop hybrid seed for a variety of reasons, and there is no clear
tendency to stay with hybrid once it is first tried.

[8] Ing. Antonio Vera Mora, of the Instituto Nacional Indigenista in San
Cristóbal, was very helpful regarding the problems discussed in this section.

some farmers, even when they were careful to find seed corn native to the area, for rainfall patterns differ in different parts of the Grijalva basin (see Helbig 1964). For instance, one farmer reported missing the optimum time for seeding during his first year of work in Zone 9.

In sum, the shifts to new areas that are described in the earlier sections of this chapter, as well as more strictly technological innovations, involve uncertainties that have produced unexpected results for some of the farmers who have dared to try something new. Given the marginal land usually farmed by Zinacantecos, the basic opportunities offered by new roads and new markets have produced more frequent and more important successes than the technological innovations, which are limited by the quality of the land. In the long run, chemical weed killers, hybrid corn, and fertilizer may become more widely used; but at present the opportunities they offer are more uncertain than the opportunities afforded by access to new land and new, stable markets.

A Measure of Increasing Prosperity

Chapter 6 shows that a farmer who works in more distant zones should earn more for the same amount of effort on his own part. Since Zinacantecos, as a group, have gradually moved to the more distant zones, it may be concluded that they are on the whole more prosperous than they were ten years ago. There are, in fact, many signs of increasing prosperity in the Zinacanteco community; but few of them offer a solid base for comparison with conditions ten years ago, and none that I know of allow the conclusion of increasing prosperity to be extended to the entire population. That is, they show that some people are better off, but not that people on the whole are better off. Although the conclusions reached in Chapter 6 clearly point toward greater income from the new farming opportunities, my data on wages (from the work histories) provide a much more certain measure of what has happened, and they allow conclusions about effects on those Zinacantecos and inhabitants of other communities who are at the bottom of the economic scale, the workers.

Table 7.9 shows the tabulation of wages paid for the years from

TABLE 7.9
Workers' Pay in Cash and Corn, 1957–66
(*from work histories*)

Year	Mean corn wage in almuds/week	Number of informants	Mean cash wage in $/week	Number of informants
1966	5.2	11	$45.20	9
1965	5.2	14	43.10	7
1964	5.1	13	42.60	7
1963	5.0	11	36.20	6
1962	4.9	11	33.60	5
1961	4.7	10	34.00	6
1960	4.8	11	33.70	6
1959	4.8	8	30.00	4
1958	4.5	6	30.00	4
1957	4.6	5	31.20	5
All years	4.9		37.50	

1957 through 1966. The increase in cash wages (almost 50 percent) has of course been eroded by the slow general inflation in Mexico; but the gains in wages paid in kind are undeniably real gains that are reflected in improved levels of living. Though the data are limited and the figures no doubt unreliable in part, they seem clear enough to support the conclusion that real wages for workers have risen at least 10 percent during the decade under study. This change has implications for the farmer who attempts to run a large operation in the traditional zones. Assuming fairly constant, if not declining, yields in these zones, his increasing labor costs mean that he is operating on a smaller margin of profit. Although it is part of the general complex of changes (and perhaps more a result than a cause), the increase in real wages may also be seen as a factor inducing farmers to move to more distant and more productive fields.

The Changing Role of the Farmer

New opportunities in physical location of work, mode of transport, and marketing are the most immediately impressive aspects of the changes Zinacanteco farmers have seen over the last decade.

However, the shifts in the farmer's role that these changes demand will probably continue to affect the society long after the immediate changes have become part of the expected pattern of life. The social relations essential to farming and the manner in which the typical farmer spends his time are rapidly changing to what is usually called a more modern pattern. Specialization and universalistic relationships are replacing the diffuse and particularistic economic and social roles that have characterized the traditional Zinacanteco peasant.

These changes are clearest in the areas of transport and marketing. The traditional use of pack animals to transport the harvest involved complicated procedures and a substantial investment of time. Few farmers ever had enough mules to bring their corn home in a single trip, or even a very few trips. Typically, a farmer had to seek out mule owners to request the use of mules on specific dates. The favor was asked and the details and price–set over a gift of drink presented to the mule owner by the farmer. And norms seem to have been those of a seller's market: that is, the owner expected to be treated delicately and expected to have his service appreciated. When the final trip was completed, the owner and any assistants were given a meal like that provided for workers at the end of a period in the lowlands. News would often be sent ahead so that the meal would be ready for the party's arrival, and some farmers signaled their arrival by blowing a horn as they came over the last major ridge on the way home.

Although these traditional practices may appear colorful to an outsider, most farmers I have talked with do not remember the rushed trips up and down the mountainside with fondness. They are tired at the end of the harvest, and enthusiastic about the use of trucks, which allows them to take a free ride home on top of their corn. Truck drivers may be fed or treated to a drink, but during harvest season they are in constant demand and rush about trying to earn as much as possible. Most of them are relatively anonymous residents of other towns, and the farmer has no continuing relations with them. From the farmer's point of view a process of transport that used to involve days of planning, a trip to

San Cristóbal to buy meat for a meal, and a week or more of work, may now take less than a day; what once required time, a payment in corn, and many continuing social relationships now involves cash.

The new marketing practices involve even more important changes in the way the farmer typically arranges his life. Prices have risen and stabilized. The rock-bottom price offered at harvest time in the lowlands, which was often less than half the peak price in the San Cristóbal market, has been replaced by the receiving centers' price, which is very close to the San Cristóbal price when transport costs are considered. At the same time, the extraordinary peaks in San Cristóbal prices that allowed some speculators to make a killing in the late summer market are less likely to appear. Formerly, only a man who was desperately in debt would sell the bulk of his crop at harvest time, either in the fields or in San Cristóbal. Today, wholesaling the crop in the lowlands is very often the best economic strategy. A farmer is now able to take home only what he needs for his family and for use as production capital in the coming year. Most farmers who sell their surplus over production capital and family needs in the lowlands still bring home a bit of corn to store for later sale in San Cristóbal; but I have met at least one man who sold his entire crop in the lowlands and bought his family's supply in San Cristóbal when the market was temporarily depressed.

In sum, the merchant roles of retailer and warehouser that were so much a part of the successful farmer's economic life ten years ago have effectively disappeared for those who choose to discard them; and the large house and mules in the yard that characterized important families are becoming less common as they become less essential to success. That is, the peasant who took corn from the seed to consumer's basket is being transformed into a specialized agricultural producer who leaves transport, warehousing, and retailing to others. None of these trends is widespread enough to permit any predictions regarding stable new configurations of social and economic activity (if that is what lies ahead), but they are all clear enough to insure that the typical Zinacanteco will soon be much less a peasant than he was a decade ago. As corn farming becomes more a specialized occupation and less a way of life, we may expect greater numbers of Zinacanteco farmers to become disaffected. Where their new and potentially distressing freedom will take them is not yet clear.

EIGHT

Stratification and Chance-Taking

MUCH HUMAN behavior involves substantial uncertainty of some kind. The actor often proceeds without being sure of the outcome of his actions in a situation where highly undesirable outcomes may be as likely as highly desirable ones. This is certainly the situation of the many Zinacantecos who adopted the new farming and marketing practices described above. They took chances, whereas those who retained traditional practices did not. In this chapter, I will present a general theory relating rank in a stratification system and inclination to take chances, and will then test it with data on Zinacanteco corn farmers and data from studies done by other investigators.

The theory presented here is of course just one of many that are relevant to predicting differential response to opportunities for economic innovation. No single independent variable, even one so powerful as rank in a stratification system, will explain all the variance in so complex a natural situation. And no single conceptualization of the dependent variable, even one so central and powerful as chance-taking, will characterize every aspect of what a farmer is doing when he changes his behavior. On the other hand, the material that follows does show the importance of noneconomic variables in decision-making under uncertainty.

Chance and Risk

In the earlier presentation of this theory (Cancian 1967) I used the term "risk" where I am now using "chance." I have used "chance" here to emphasize that the principal uncertainty referred to is the unmeasurable type existing in situations where lack of knowledge is an important factor. That is, I have tried to follow Knight's distinction (see Chapter 1) by using the word chance to describe action taken under uncertainty resulting from substantial lack of knowledge, reserving the word risk for the aspect of uncertainty that involves outcomes with known or fairly certain long-run probabilities of less than one.

Since actors make a single observable decision, not one decision in terms of "chance uncertainty" and another in terms of "risk uncertainty," the distinction is strictly an analytical one. In terms of the theory that follows, chance-taking is the part of behavior that can be explained in terms of noneconomic variables—specifically rank in a stratification system. The part of behavior explained in terms of economic ability (i.e., wealth as a facilitator of expensive action) and differential knowledge, which might more properly be included in a microeconomic model, is risk-taking. Given the data I have, the distinction between chance-taking and risk-taking is eventually expressed in terms of the different explanatory variables: rank, economic ability, and knowledge. However, since both chance-taking and risk-taking are involved in the observed behavior, the use of either single term to describe this behavior leaves something to be desired.

Although I will be trying to isolate behavior that can be explained by noneconomic variables, I have chosen to use the term "risk" rather than "chance" in the remainder of this chapter. In large part, this decision is based on the fact that "chance" is a less flexible English word than "risk." "Chance-taking," "inclination to take a chance," and "degree of chance" are simply more awkward than "risk-taking," "inclination to risk," and "degree of risk."[1] In

[1] The awkwardness of "chance" words makes me think that the failure of economics to advance substantially in the area opened by Knight's distinction is in part due to simple problems of expression. The current surge in importance of Bayesian statistics in economics may open the way for the inclusion of different states of actors in economic analysis, for it is possible to include a variety of noneconomic factors in the actor's subjective probability estimate. The Bayesian viewpoint, of course, does not automatically provide knowledge about the complex noneconomic variables necessarily involved in such a reorientation.

any case, the validity of the analytical distinction between chance
and risk hangs not on the word chosen to describe behavior but
on the success of the attempt to explain decisions about economic
activities in terms of noneconomic variables, such as rank.

The Theory of Stratification and Risk-Taking

At the outset, I will assume that it is in the nature of stratification systems for any individual to prefer high rank to low rank.
Most of the theory flows from this proposition and from the idea
that the possibility of achieving higher rank is often the motivation
for risk-taking. Rank will be conceptualized as the possession or
control of resources, and risk-taking as the deployment of resources
in situations of uncertainty. The relation of rank and risk-taking
that I eventually put forward is curvilinear rather than linear, and
my exposition goes through a two-stage process in which the second stage presents apparent contradictions of statements made in
the first stage. Despite the disadvantages of such an expository se-
quence, this is still the best way I have been able to find to present
the theory.[2]

Although the stated focus of the theory is risk-taking, it may be
useful to point out that a very common and interesting special case
of risk-taking is found in innovative behavior; and, further, that
innovative behavior is often contrasted with conservative behavior.
Thus, although I seek to explain who will take risks, the theory
may at the same time be seen as explaining who is conservative,
i.e., who will not risk.

The Inhibiting Effect of Relatively High Rank

Rank in a stratification system is the possession or control of
resources valued by the society in which the system exists. Since
it is assumed that in all societies individuals would rather be
higher than lower on any ranking,[3] the problem is to predict how

[2] The earlier statement of the theory (Cancian 1967) employed the same se-
quence.

[3] Obviously, actors in natural situations are involved in a number of different ranking systems at one time; and they are often required to economize

they will manage their resources in their efforts to achieve the highest possible rank. In general, I think, it is clear that the higher a person's rank the more disadvantageous to him a random change in rank is apt to be; that is, the higher a person's rank the more likely it is that a random change will be downward rather than upward. Persons of higher rank will therefore be less enthusiastic about random changes of rank than persons of lower rank. Thus, if we see risk as the random element in situations that may affect rank, we are led to the conclusion that *persons of higher rank will risk less than persons of lower rank.* This statement of what I call the inhibiting effect of high rank (Figure 8.1, Curve *I*) depends on the assumption that all other things are equal, and thus involves many implicit assumptions about natural situations that are in fact untenable. All other things are not equal; many other characteristics that are closely tied to rank on most important ranking dimensions must be controlled if this proposition is to be a useful predictor of human behavior. The proposition stated above might be expected to hold with the following definitions and conditions:

DEFINITION 1. *Risk is a characteristic of situations of exchange in which the rate of return on investment of resources is uncertain; the greater the uncertainty, the greater the risk.*

DEFINITION 2. *Equal risk, for persons of different ranks, is represented by the investment of an equal proportion of the total resources of each person under conditions of equal uncertainty.*[4]

CONDITION 1. *All risks (uncertain investments of resources) are perfectly divisible.*

---

between their positions on different hierarchies. Because of this, any statement of a theory referring, like this one, to a single resource involves a massive assumption that other factors are constant. In practical research situations the theory cannot be expected to apply unless: (1) it is possible to control other factors so that the assumption is realistic; or (2) the resources under study are so important to so many people that the predicted effects may be seen despite the uncontrolled variance of other resources.

[4] It may be more appropriate to say that equal risk is represented by the investment, under conditions of equal uncertainty, of an equal proportion of the resources that separate each person from the next lower category in the hierarchy. However, the statement in the text avoids complications of measurement.

CONDITION 2. *Knowledge is equally spread over all ranks.*

CONDITION 3. *The risk necessary to maintain present rank is equal, as a proportion of total resources, for persons of all ranks.*

CONDITION 4. *No individual can suffer total loss of resources from loss on a single risk.*

CONDITION 5. *No individual has so many more resources than the next lower relevant individual (or category of individuals) that he is completely protected from loss of rank.*

Clearly, the conditions stated above are not met in most natural situations; hence the theory must be modified before it can be applied and tested. As I see them, the necessary modifications are two. The first (involving Conditions 1–2) is a rather straightforward adjustment to take account of the fact that risks are not perfectly divisible, and of the fact that high rank in most resource hierarchies tends to bring its possessor a relatively large amount of relevant knowledge. The second modification (involving Conditions 3–5) is much more fundamental. It calls into question the applicability of the initial proposition about the relation of rank and inclination to risk to individuals who occupy the very highest and very lowest ranks. The two modifications are discussed in order below.

The Facilitating Effect of Relatively High Rank

If the conditions stated above are met, the relationship of resource rank and inclination to risk would be linear and negative, and would be readily apparent in natural situations. However, the conditions are not met in most natural situations. As a result, high-ranking persons appear to act in high-risk ways more often than low-ranking persons, even though the inclination to risk of the high-ranking persons may be less than that of lower-ranking persons.

If Condition 1 ("All risks are perfectly divisible") is not met, the low-ranking person is much more likely to be at a disadvantage when it comes to behaving in what appears to be a risk-taking way. A poor farmer, for example, may be much more inclined than a rich one to risk the uncertainties of a new type of fertilizer; but one bag of it may represent a large investment to him, whereas it is a small investment and a small risk to a richer man. If we look at risk-taking in a situation where the fertilizer is being introduced, we may see the poor man as very conservative and unwilling to take risks, when by any reasonable standard he is not.

The second condition ("Knowledge is equally spread over all ranks") is likewise not met in natural situations, and its absence normally has an effect similar to that just mentioned. Relevant knowledge is usually spread unevenly among the members of any stratification system, and high-ranking members have more knowledge than low-ranking members. Since the risk-taking that concerns us here is largely a product of uncertainty, this unequal distribution of knowledge may affect the behavior of individuals by removing relatively larger amounts of uncertainty for those higher in rank. As a result, they may appear to be taking risks when their actual inclination to risk is in fact less than that of lower-ranking individuals who have less knowledge.

Thus identical acts, from an outsider's point of view, involve different actual or subjective risk for persons at different points in the rank hierarchy. The higher-ranking individual often takes less actual risk than the lower-ranking individual, although they seem to perform the same act defined in terms of an outside point of view that does not consider the rank of the actor or, by extension, his postulated internal state. In the study of risk-taking in natural situations, it is useful to recognize this overall tendency, which is summarized in the label "the facilitating effect of high rank" and illustrated by Curve F in Figure 8.1.

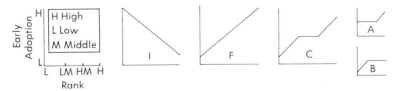

FIGURE 8.1

The Ends of the Ranking Continuum: The Curvilinear Effect

If we see a person's inclination to risk as a balancing of what he has to gain and what he has to lose, it is not difficult to imagine that people at the ends of the rank continuum might not operate according to the principles that predict the behavior of those in the middle of the continuum. Those at the end have everything or nothing to gain, everything or nothing to lose. When this is the case, one might argue, an economizing conceptualization of the situation does not make sense; those who define the ends of ranking continua may not participate in the competition for rank in the same way as those who are not so conspicuous. Of course, the question of how they will behave still remains open. I argue that they will behave *unlike* the prediction made for them by the linear principles discussed above. The behavior most unlike the predictions of the initial proposition would be for the highest-ranking people to be high riskers and the lowest-ranking people low riskers. These statements yield a relation of resource rank and risk-taking like the curvilinear one shown in Figure 8.1, Curve *C*. Note that in order to illustrate this idea a minimum of four ranks must be distinguished. For convenience, I will refer to these ranks as Low, Low Middle, High Middle, and High.

The kind of thinking that goes into arguments for the distinctiveness of people at the ends of the rank continuum is illustrated by George Homans's chapter on "Status, Conformity, and Innovation" (Homans 1961). Homans usually divides the social continuum into three parts: high, middle, and low. He argues that highs are innovative because they can afford it or because they need to maintain their distinctiveness. Lows are innovative because they are not highly rewarded by the group for any of their behavior; thus they find the cost of nonconformity to be more consistent with their expectations than do the middles, who are accustomed to rewards for their conventional behavior. This produces a U-shaped curve.

Homans finds a different interpretation necessary to handle the results of an experiment by Dittes and Kelley. In their work, lows and very-lows were distinguished, making four ranks in all; and the very-lows were extremely conformist in their publicly expressed opinions. Here, Homans accepts the interpretation made by Dittes and Kelley: "In the extreme case, however, where acceptance is so low that actual rejection is presumably an imminent possibility, anxiety about rejection is especially high, and the result seems to be a pattern of guarded public behavior" (Dittes and Kelley 1956: 106).

In both of Homans's arguments, "middle-class conservatism" is explained by focusing attention on the other "classes" and giving reasons for their innovative spirit or their ultra-conservatism. And in both, although similar variables are considered throughout, essentially separate arguments are presented for each end of the continuum. The argument that the ends of a continuum will be unlike the middle of the continuum simply because of their nature as extremes, which was presented casually at the beginning of this section, is the only sociologically oriented argument I know that does not involve separate conceptualizations of what is going on at each end of the continuum. But it is hardly an argument at all, for it merely says that the ends of a rank continuum are always different from the middle.[5]

There are at least two approaches that yield the conclusion I have suggested (i.e., that shown in Figure 8.1, Curve C). They both make separate arguments for each end of the ranking continuum: i.e., one argument that makes the highest group more innovative than otherwise expected (the "A" effect, represented by the A curve in Figure 8.1); and another argument that makes the lowest group less innovative than otherwise expected (the "B" effect, rep-

[5] Wilkening, Gartell, and Presser (1969), in a paper read before the Rural Sociological Society, have pointed out that the apparent special effects predicted for the ends of the continuum may in large part be seen as the results of a combination of a normal distribution and a positive function. Assume that inclination to risk is normally distributed (i.e., forms a bell curve). Assume that it is positively related to wealth. Break the normal distribution into quartiles and plot them as I have in Figure 8.1. The result will have the characteristics of the $F > I$ curve in Figure 8.2 (see p. 144). This explanation has the virtue of simplicity, but it cannot produce the negative relation shown in the $I > F$ curve in Figure 8.2. This latter relation is, of course, at the heart of the present theory.

resented by the *B* curve in Figure 8.1). The first approach sees the actors at the ends of the continuum as motivated to compete for the highest possible rank; but it sets forth special conditions applying at the ends of the continuum that make the behavior of these actors different from that predicted by the initial proposition. The second approach interprets the actors' positions as the basis for noninvolvement in the competition for rank. The further details of the arguments are idiosyncratic.

Argument 1A. The rich realize that their distinctiveness is based on leadership in economic techniques and take calculated risks in order to maintain this distinctiveness. If this argument is correct, Condition 3 of the original theory does not hold for the High rank.

Argument 1B. The poor are so poor that any risk threatens total economic extinction; therefore, they are unusually conservative (the Dittes and Kelley point). If this argument is correct, Condition 4 of the original theory does not hold for the Low rank.

Argument 2A. The rich are secure in their high position and take flyers because they have little to lose by doing so. If this argument is correct, Condition 5 of the original theory does not hold for the High rank.

Argument 2B. The poor refuse to compete in the economic sphere because past failures have made it seem an inefficient way to seek rewards. This statement may be taken as a parallel to Homans's argument (with a three-class division) that lower-class people will be nonconformists (innovative) because their attachment to the group and its rules is low. In reference to a social group, this means that they will do unusual things, and that these things will be encoded on the social dimension and will make them nonconformists or innovators. In a situation of economic innovation, however, where the relevant resource (wealth) is more strictly defined than it can be in a social group, their nonconformist behavior may take place entirely outside the economic domain. Given his low economic success in the past, a person of Low rank has a low attachment to the economic system and may "innovate" by spending his few resources outside the sphere of production. This statement simply attempts to explain the folk observation that poor people waste their money on foolish things. Since they do not attempt to compete in the economic sphere, they will be especially low innovators. If this argument is correct, the principal proposition of the theory is itself not true for the economic behavior of the Low rank.

These arguments for the curvilinear effect may be seen as direct revisions of the initial proposition relating resource rank and risk-taking—i.e., "Individuals of higher rank will risk less than individuals of lower rank." This proposition might still hold if all the conditions under which it was stated could be found in a natural situation or made to exist in an experimental situation; but more likely than not, they will not exist in the natural situations that are my primary concern here. Thus modifications of the theory that contradict its five conditions must be incorporated into the theory and used directly in making predictions from it. For our present purposes this presentation of the theory may be seen as an argument that the initial proposition holds for the middle ranks only, and that other principles must be used to predict the behavior of other ranks. But if such a viewpoint on the theory had been taken from the outset, the result would have been a more fragmented and less general theory. The presentation that I have used extends the theory to situations where the conditions might hold, though I do not expect to find any in the natural situations that concern me in this study.

I will now develop hypotheses that permit the testing of this overall theory of rank and risk-taking.

Predictions from the Overall Theory

Insofar as the inhibiting, facilitating, and curvilinear effects present an adequate overall theory of the relation of rank and risk-taking, any curve representing an empirical case of this relation must be recognizable as a combination of the I, F, and C curves. Since Curves I and F negate each other, the inhibiting effect and the facilitating effect will never be evident in the same descriptive curve, although they may well be present in the same empirical situation. Any empirical curve must be either a C curve (when I and F exactly negate each other) or a combination of I or F with C (see Figure 8.2).

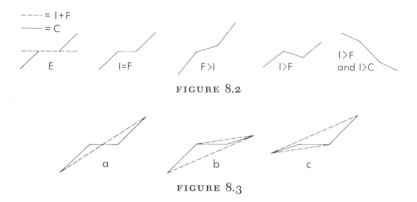

FIGURE 8.2

FIGURE 8.3

In the absence of any knowledge about the dominance of either the inhibiting effect or the facilitating effect, the distinctive features contributed by the curvilinear effect are the only characteristics of an empirical curve predictable from the overall theory. Whatever the other conditions, if there is a curvilinear effect (see Figure 8.3a), then:

HYPOTHESIS 1: *Individuals of High Middle rank will risk less than would be predicted by an assumption of linearity from Low rank to High rank.*

HYPOTHESIS 2: *Individuals of Low Middle rank will risk more than would be predicted by an assumption of linearity from Low rank to High rank.*

These hypotheses frame the predictions based on the curvilinear effect in their strongest form. Although this strong statement is desirable, it may be unrealistic for tests involving crudely analyzed data. In the discussion of the basis for predicting the curvilinear effect, two separate sets of arguments were made: one for the especially high risk-taking of the High rank (the "A" effect), and one for the especially low risk-taking of the Low rank (the "B" effect). Unless these arguments apply with equal strength (making the C curve "symmetrical"), the hypotheses stated above might be rejected (see Figure 8.3).

A weaker hypothesis that retains the prediction about the relative conservatism of the "middle class" (here, the High Middle rank) may be formulated. To retain the prediction and eliminate the possibility that the pattern will be obscured by an "asymmetrical" curve, Hypothesis 1 may be restated as

HYPOTHESIS 3: *Individuals of High Middle rank will risk less than would be predicted by an assumption of linearity from Low Middle rank to High rank* (Figure 8.3b).

In parallel fashion, the prediction that the Low Middle rank will risk more than would be expected may be protected from an "asymmetrical" curve by restating Hypothesis 2 as

HYPOTHESIS 4: *Individuals of Low Middle rank will risk more than would be predicted by an assumption of linearity from Low rank to High Middle rank* (Figure 8.3c).

Although this hypothesis predicts the behavior on the Low Middle rank, its confirmation depends heavily on the phenomenon of central interest, the conservatism of the High Middle rank.

The overall theory makes no prediction concerning the dominance of the inhibiting effect or the facilitating effect. If neither of these effects dominated the other, the curvilinear effect would dominate, producing a roughly positive relationship between wealth and risk-taking; and such a result is often reported (see Rodgers 1962). However, when the wealth continuum is divided into four parts, as it is in this study, the curvilinear effect by itself predicts no difference between the Low Middle rank and the High Middle rank. Thus dominance of either the inhibiting or the facilitating effect in an empirical situation should show up in the relationship of these two ranks: if the inhibiting effect is dominant, the Low Middle rank should risk more than the High Middle rank; and if the facilitating effect is dominant, the reverse should be true. Since the dominance of the facilitating effect would confirm the most commonly reported finding, a hypothesis may be stated:

HYPOTHESIS 5: *Individuals of High Middle rank will risk more than individuals of Low Middle rank.*

Since the theory developed here gives no reason to expect otherwise, *the prediction is that Hypothesis 5 will not be confirmed.*

If Hypothesis 5 is not confirmed, three explanations are possible. First, both the inhibiting and the facilitating effects may be operative and of roughly equal influence. Second, neither the inhibiting nor the facilitating effect may be operative, so that the curvilinear effect is completely responsible for the relationship of wealth to risk-taking. (In either of these two situations, the commonly reported finding represented by dominance of the facilitating effect would be suspect.) Third, the dominance of the inhibiting effect may be such as to confirm the hypothesis that individuals of Low Middle rank will risk more than individuals of High Middle rank. This explanation would be startling, for it would be a direct contradiction of established findings.

Predictions by Degree of Risk: The Inhibiting Effect

Since the inhibiting effect is dependent on the risky aspect of actions for which predictions are being made, there should be a variance in its influence on outcomes as the riskiness of actions varies. If comparable situations that differ only in the riskiness of some action are available, we should be able to see, for example, that Hypothesis 5 is more often confirmed in less risky situations.

The hypotheses above are stated for the early stages of the innovation process. In such a process, risk goes down as time passes and knowledge about the innovation spreads. Thus, the innovation process may be seen as having at least two stages: Stage 1 in which inclination to risk is important; and Stage 2 in which inclination to risk is substantially less important. Since the overall theory is heavily dependent on inclination to risk, it is principally applicable to Stage 1.

Here, the hypotheses about the difference between Stage 1 and Stage 2 situations will be stated in general terms. The first hypothesis cannot be applied to all pairs of ranks. Because of the indeterminate influence of the curvilinear effect, it must be limited to comparisons of the High Middle rank and the Low Middle rank.[6]

[6] For comparisons of the High rank with one of the three lower ranks, the curvilinear effect may produce dominance of the High rank over the lower

HYPOTHESIS 6: *The risking rate of the High Middle rank in low-risk situations (Stage 2) minus its risking rate in high-risk situations (Stage 1) will be greater than the risking rate of the Low Middle rank in low-risk situations minus that rank's risking rate in high-risk situations.*

Predictions by Degree of Risk: The Curvilinear Effect

In the above exposition of the theory, two sets of arguments were made for the curvilinear effect. One set is based on attributing to the High and Low ranks idiosyncratic economic (rank-seeking) motives for risking and not risking. The other involves noneconomic motivation for economic behavior. If the first set of arguments is correct, then the curvilinear effect should lessen in low-risk situations. If the second is correct, the curvilinear effect should not lessen in low-risk situations.[7] Thus we have a way of evaluating the two sets of arguments.

In the discussion preceding the presentation of Hypothesis 6, it was observed that both the inhibiting and the curvilinear effects may be expected to change from high-risk to low-risk situations. Thus any test of change in the curvilinear effect must control change that is due to attenuation of the inhibiting effect. To do

---

rank in a high-risk situation. If the curvilinear effect lessens in low-risk situations, the risk-taking of a lower rank may increase more than that of the High rank because the gain to the lower rank from the lessening of the inhibiting effect may be more than the gain to the High rank from the lessening of the inhibiting effect minus the loss to the High rank from the lessening of the curvilinear effect. A similar argument applies to comparisons of the Low rank with the three higher ranks. The middle ranks remain unaffected by the indeterminacy of the curvilinear effect and are therefore useful for testing the hypothesis about the relative strengths of the inhibiting and facilitating effects.

[7] Although the implications of Argument 2B for change or lack of change from high-risk situations to low-risk situations present no problem, the implications of Argument 2A are not so clear. Argument 2A states that the rich are secure in their high position and take flyers because they have little to lose by doing so. This argument is ambiguous at best; and it may be that the attractiveness of new opportunities as "flyers" is lower in low-risk situations than in high-risk situations. If this is true, Argument 2A could be used to predict a reduction in the risking rate of the High rank, thus making Arguments 1A and 2A indistinguishable. For the present purposes, I assume that new opportunities do not lose their attractiveness as "flyers" so quickly. Thus, as implied in the text, Arguments 1A and 2A lead to different predictions about change or lack of change in the adoption rate of the High rank from high-risk situations to low-risk situations.

this, expected values for the High and Low ranks may be calculated by extrapolation from the line defined by the Low Middle and High Middle ranks, which are not influenced by the curvilinear effect (see Curve *E* in Figure 8.2). If the curvilinear effect lessens in low-risk situations, then deviation of the Low and High ranks from the expected values defined in terms of the line from Low Middle to High Middle rank should be less. Separate predictions for the lessening of the "*A*" effect (High rank) and "*B*" effect (Low rank) components of the curvilinear effect may be framed as follows.

HYPOTHESIS 7: *The positive deviation of the High rank from the expected values defined by the line from Low Middle rank to High Middle rank will be less in low-risk situations than in high-risk situations.*

HYPOTHESIS 8: *The negative deviation of the Low rank from the expected values defined by the line from Low Middle rank to High Middle rank will be less in low-risk situations than in high-risk situations.*

Testing the Stratification and Risk-Taking Theory

For these tests, risk-taking will be operationalized in terms of the behavior of Zinacanteco farmers in the face of the changes described in Chapter 7. The two major changes are movement to more distant fields and use of the government receiving centers for marketing. At the outset, each of these changes represented a substantial risk. For the purpose of testing the hypotheses, I will assume that the more distant a farmer's fields in 1966, the greater the risk he was taking relative to his fellows. In the case of the receiving centers time rather than distance provides the necessary variance. It is assumed that the earlier the farmer made his first use of the receiving centers (in terms of years), the greater the risk he took relative to his fellows.

A farmer's rank on the continuum will be measured by the amount of corn he seeded in the lowlands during the 1966 crop year. Since the theory explicitly limits itself to predictions about behavior relevant to the particular resources that are being risked, amount of corn seeded rather than overall wealth seems to be the most appropriate measure of rank.[8] That is, a farmer's behavior in lowland farming should reflect the investment he has in that farming. Clearly, if he obtains just a small part of his overall wealth from lowland farming, his behavior in handling his investment may be influenced to the degree that his gain from lowland farming is more or less important to his total economic life.

Treatment of Data

The hypotheses require that the rank variable be divided into four parts. Since the distribution of Zinacantecos on the measure of rank (corn seeded in the lowlands) is not continuous, the most straightforward procedure for approximating quartiles was adopted: the distribution was divided into two parts that were as equal as possible given the discontinuities in the distribution; and each "half" was then divided similarly. As can be seen in the tables below, the dividing points are different for Nachig and Apas, and for the two Nachig populations. The difference between the hamlets comes from the fact that Nachig farmers seeded more than Apas farmers (mean almuds seeded: Nachig = 3.5, and Apas = 2.9).[9] Almost half the Nachig farmers (44 percent) seeded more than three almuds, whereas the comparable figure for the Apas population is 25 percent. This mechanical procedure for dividing the populations into quartiles seems best for the formal testing of hypotheses, for it insures against arbitrary manipulation of the data.

Following a procedure used in the tests of the theory completed before this study was done (see Cancian 1967), the risk-taking variables were divided by taking the best approximation to the first quartile (highest riskers, or Stage 1 adopters), then the best approximation to the second quartile (Stage 2 adopters), leaving

[8] Corn seeded in the lowlands and overall wealth are in any case highly correlated: for Nachig, Kendall's Tau = .70; for Apas, Kendall's Tau = .59.

[9] This difference may be due at least in part to differences in informants. I have no good way of checking the reports of the informants. My impression is that the difference is a real one, and that little informant bias is included in these figures.

the remainder of the population undivided. Since the high riskers are the ones that most concern us here, this procedure was adopted to best approximate comparable groups across populations in the face of the discontinuities in the data.

The data, divided as discussed above, are displayed in Table 8.1. As can be seen, discontinuities in the original data make it impossible to get very good approximations to quartiles. To make these

TABLE 8.1

Cross Classification of Farmers by Rank and
Degree of Risk-Taking

(*from survey*)

Population and degree of risk-taking	Rank (by almuds seeded in lowlands)[a]				
	Low	Low Middle	High Middle	High	Total
Nachig location:					
Stage 1 (Zones 8, 9)	4	22	12	15	53
Stage 2 (Zones 6, 7)	5	12	7	7	31
Other (Zones 2, 4, 5)	23	43	28	15	109
TOTAL	32	77	47	37	193
Apas location:					
Stage 1 (Zone 6)	2	5	1	11	19
Stage 2 (Zone 3)	0	4	11	5	20
Other (Zone 2)	16	23	19	11	69
TOTAL	18	32	31	27	108
Nachig marketing:					
Stage 1 (1962–63)[b]	7	11	15	17	50
Stage 2 (1964–65)	8	9	11	10	38
Other (1966, 0)	41	6	16	8	71
TOTAL	56	26	42	35	159
Apas marketing:					
Stage 1 (1962–63)	2	7	5	11	25
Stage 2 (1964–66)	3	3	7	7	20
Other (0)	9	14	14	5	42
TOTAL	14	24	26	23	87

[a] Almuds seeded by rank varied between categories as follows. Nachig location: Low, 1; Low Middle, 2–3; High Middle, 4–5; High, 6–8. Nachig marketing: Low, 1–2; Low Middle, 3; High Middle, 4–5; High, 6–8. Apas location and marketing: Low, 1; Low Middle, 2; High Middle, 3; High, 4–8.

[b] These are the years in which farmers in the corresponding row began selling to the receiving centers; 0 means they have never sold corn there. The populations for marketing ($N = 159$, $N = 87$) include only farmers who worked independently in the lowlands for all of the five-year period used to rate their inclination to risk.

TABLE 8.2
Risk-Taking and Rank (Standardized Percentages)
(*from Table 8.1*)

Population and degree of risk-taking	Rank			
	Low	Low Middle	High Middle	High
Stage 1:				
Nachig location	12%	27%	24%	38%
Apas location	16	22	5	58
Nachig marketing	9	30	26	35
Apas marketing	13	26	17	43
Stage 2:				
Nachig location	19	24	22	35
Apas location	0	18	44	38
Nachig marketing	9	35	24	32
Apas marketing	19	13	25	43

figures more easily comparable across populations, they were converted to percentages of columns, and then each row was converted to a standard base (100). These "standardized" figures are presented in Table 8.2, and are used to draw the illustrative curves in Figure 8.4.

Testing the Hypotheses

Conversion of the hypotheses into quantitative form that will permit testing is straightforward. For example, in Hypothesis 1 (High Middle individuals will risk less than would be predicted by an assumption of linearity from Low rank to High rank), High Middle rank is calculated by assuming that it should fall two-thirds of the way from Low rank to High rank.[10] Thus the hypothesis is:

HYPOTHESIS 1: $HM < 2/3(H - L) + L$

Quantitative statements of the other hypotheses are as follows (subscripts indicate stages, i.e., degree of risk).

[10] Although the assumption of equal intervals between ranks is required for the formulation of the hypotheses below, the spirit of the argument and the quality of the data do not require so strong an assumption. I have retained the linear assumption in order to permit the quantitative testing of these hypotheses.

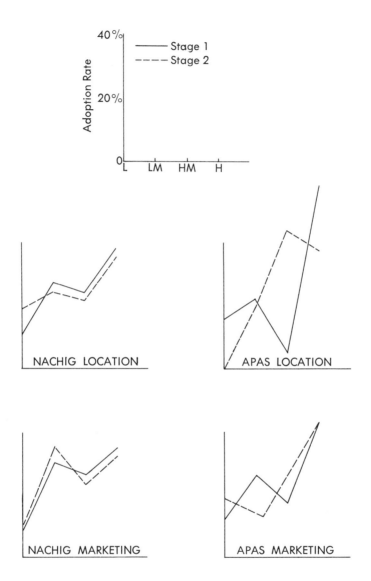

FIGURE 8.4

HYPOTHESIS 2: $LM > \frac{1}{3} (H - L) + L$

HYPOTHESIS 3: $HM < \frac{1}{2} (H - LM) + LM$

HYPOTHESIS 4: $LM > \frac{1}{2} (HM - L) + L$

HYPOTHESIS 5: $HM > LM$

HYPOTHESIS 6: $HM_2 - HM_1 > LM_2 - LM_1$

HYPOTHESIS 7: $H_2 - [HM + (HM - LM)]_2 < H_1 - [HM + (HM - LM)]_1$

HYPOTHESIS 8: $L_2 - [LM - (HM - LM)]_2 > L_1 - [LM - (HM - LM)]_1$

Results

The results of the tests are presented in Table 8.3.[11] On the whole the hypotheses are confirmed. Within Zinacantan, distinctions of economic rank have proved to be important predictors of innovative behavior. And the distinctions that are relevant are not the grossest sort of rich-poor division but a more complex and subtle sort that involves four categories of rank and a curvilinear relation of rank and risk-taking. Although the typical farmer in Zinacantan has an income of roughly $100 (U.S.) after providing corn to feed his family, and the unusually wealthy farmer may reach $400 at the most, these differences are sufficient to support an internal stratification system that produces clear differential response to opportunities for innovative behavior.

In the context of the present study these results are important because they illustrate the connection between uncertainty and noneconomic explanations of economic decision-making and because they show that Zinacanteco behavior is so clearly patterned in terms of a general theory of stratification and risk-taking.

[11] Note that neither measure (location or marketing) on either population (Nachig or Apas) confirms Hypothesis 5. All four instances would confirm the hypothesis that the inhibiting effect is dominant. This clear (if limited) pattern of support for the initial proposition is not explained by the Wilkening, Gartell, and Presser reinterpretation of the basis for the curvilinear effect (see note 5). Thus I am encouraged to maintain the overall theory intact despite their important criticism.

TABLE 8.3
Results of Tests

Category	Hypothesis							
	1	2	3	4	5	6	7	8
Nachig location	X	X	X	X	O	X	X	X
Apas location	X	O	X	X	O	X	X	X
Nachig marketing	X	X	X	X	O	O	O	O
Apas marketing	X	X	X	X	O	X	X	X

NOTE: X indicates that the hypothesis was confirmed, O that it was not confirmed. The prediction for Hypothesis 5 was that it would not be confirmed.

Further Support for the Theory

Before the major field work for this study in 1966–67, a comparative study was done (Cancian 1967) using data on the use of receiving centers by Apas farmers (through 1965) and data from six studies done by other investigators. Five of the other studies concerned agricultural populations in the United States, and the sixth (Lindstrom 1958) was done in Japan. Table 8.4 shows the measures of wealth and adoption rate for all seven studies; Figure 8.5 displays the data in graph form parallel to Figure 8.4; and

TABLE 8.4
Measures of Wealth and Adoption Rate, by Study

Study	Measure of wealth	Measure of adoption rate
Cancian (1967)	Corn (maize) seeded	Year of first sale to receiving center
Dean *et al.* (1958)	Acres in crops and improved pasture	Corn practice score (an index)
Fliegel (1957)	Net farm income	Twenty recommended practices (an index)
Gross (1942)	Net income	Year of first use of hybrid corn seed
Lindstrom (1958)	Farm size	Four recommended practices (an index)
Marsh and Coleman (1955)	Acreage operated	Year of first use of bluestone lime
Wilkening (1952)	Acres of cropland operated	Improved farm practices (an index)

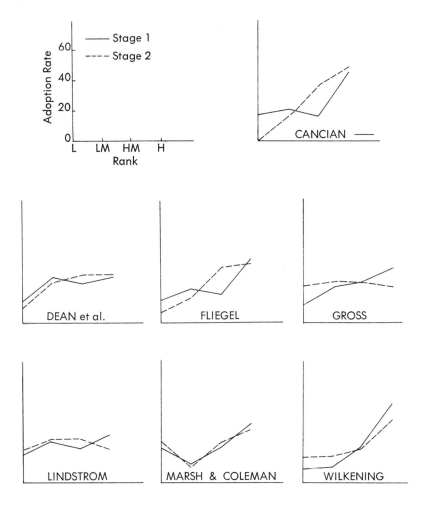

FIGURE 8.5

TABLE 8.5
Results of Tests by Study

Study	Hypothesis							
	1	2	3	4	5	6	7	8
Cancian (1967)	X	O	X	X	O	X	X	X
Dean et al. (1958)	O	X	X	X	O	X	X	X
Fliegel (1957)	X	O	X	X	O	X	X	X
Gross (1942)	X	X	X	X	X	O	X	X
Lindstrom (1958)	X	X	X	X	O	X	X	X
Marsh and Coleman (1955)	X	O	X	O	X	X	X	X
Wilkening (1952)	X	O	X	O	X	O	X	O
Number confirming	6	3	7	5	3	5	7	6

NOTE: X indicates that the hypothesis was confirmed, O that it was not confirmed. The prediction for Hypothesis 5 was that it would not be confirmed.

Table 8.5 shows the results of testing Hypotheses 1–8 with these data. (Details of the data and full credits to the investigators who generously contributed them are found in Cancian 1967.)

Clearly, the theory, which was developed for my 1965 study of Zinacantan and confirmed by the present more detailed study, has broader applicability. Though the total support remains scanty by most standards, I think that it justifies the speculations that follow immediately below. (The more general implications of various aspects of the theory are discussed in Cancian 1967.)

Uncertainty and the Middle Class

The major import of this study's findings lies in the downward dip of the curve between the Low Middle and High Middle ranks (i.e., the confirmation of Hypotheses 1 and 3, and the predicted failure to confirm Hypothesis 5) and in the apparent upswing of that part of the curve during the second stage of the process of change (i.e., the confirmation of Hypothesis 6). In general terms, the High Middle rank may be identified with the middle class in any society—with those who are well established in the ranking system but hold neither top rank nor any rank that could be seen as low. The present findings suggest that this group is especially conservative during the early stages of change, when uncertainty is greatest; and that it tends to lose this distinctive quality as change proceeds and the advantages and disadvantages of new practices become more apparent.

A model of change that stresses the diffusion of information might identify the middle class as the group most likely to innovate: it typically has high educational and technical qualifications; its behavior patterns approximate the societal ideal; and it apparently does not have the special interests and views that may limit groups at the ends of the rank continuum. A rational, information-oriented program of change might logically be focused on this part of the population in the hope that it would provide leadership in change.

However, insofar as the findings of the present study are correct, the middle class is a poor responder to changes that involve any substantial uncertainty. Whatever the ultimate value of this conservative tendency to the society as a whole, it must be recognized by those who have decided to promote change. Even change that draws its legitimacy from some kind of societal consensus can be frustrated if it is introduced where reluctance to act under uncertainty is the greatest.

Broader Implications of the Study

This study has shown that uncertainty and individual reactions to it are important aspects of behavior in situations of economic change. Ignorance, rather than information, is basic; hence theories of information diffusion and the framework of microeconomic theory in which they implicitly operate are of limited usefulness in understanding such change. Here, I want to briefly elaborate this conclusion and identify it with a kindred view of economic decision-making.[12]

When change affects a situation, men will grope until they

----

[12] I would also like to stress that Zinacanteco behavior has been viewed as just one more instance of general human behavior and not as Maya behavior, Indian behavior, or peasant behavior. Though it is a banal commonplace to many social scientists, this perspective needs support and emphasis within the context of peasant studies. Both comparative peasant studies and general comparisons that contrast peasants with preagricultural or industrial men pro-

acquire the knowledge and experience that will permit them to act with what is recognized as a normal or stable degree of confidence. However, new changes are often upon them before old ones are absorbed. Thus behavior is characterized not by the presence or absence of uncertainty but by degrees of uncertainty ranging from dramatic and unusual to "normal." In fact, although a simplified picture of decision-making assumes that there are situations in which certainty is, let us say, so great that uncertainty is trivial, many natural situations are so complex that certainty is never that great.

In this sense, my characterization of behavior under uncertainty in situations of change is related to Herbert A. Simon's "principle of bounded rationality." Simon's principle is: "The capacity of the human mind for formulating and solving complex problems is very small compared with the size of the problems whose solution is required for objectively rational behavior in the real world—or even for a reasonable approximation to such objective rationality" (1957: 198).[13] Any behavior that is not an explicitly intended repetition of yesterday's behavior is subject to this principle. And the principle's applicability should be most evident in situations that are recognized as involving newness and change.

---

duce legitimate and important results. However, these traditions, which see peasants as peasants, may also encourage a stereotyping that has practical effects on the relationship of peasants and nonpeasants and on research in nonanthropological disciplines newly concerned with peasants. A sophisticated view of peasant-oriented research takes its findings as far as they go in explaining the behavior of people who are peasants and looks elsewhere for matters left unexplained; but an understandably more common view seizes on the label "peasant" as a full description of the people who bear it. My impression is that researchers (and many others) presently give too much weight to "peasantness" as an explanation of human behavior, both in the sense that they treat peasant societies as relatively homogeneous and in the sense that they see peasants as substantially different from other people. I hope that more research viewing peasants in terms of general theories of human behavior will help to correct this imbalance.

[13] Simon's principle can be restated in traditional economic language by saying that the cost of information is exceedingly high. But this restatement is made in terms of economic theory without regard for the demands of empirical reality, and Simon's principle is fundamentally an empirical generalization. Uncertainty may indeed be seen as a state in which the cost of information is exceedingly high, but this fact does not reduce the importance of principles that predict action under uncertainty.

In his discussion of "rational choice in the face of uncertainty" (1957: 203–4), Simon rejects the idea that decisions under uncertainty are best made on the basis of some estimate of probability distributions in the future. He favors a model based on corrective action taken by the actor as the situation develops—i.e., on a feedback relation between the actor and the world. Although Simon's choice of a feedback model is appropriate to his concern with the characterization of actors in organizations, this choice avoids the problem of characterizing decisive decisions. Even though most administrators may make continuing interactive decisions based on feedback, some decisions (such as those made by Zinacantecos in changing locations) necessarily have a more decisive character: they commit the actor to an extended course of action that cannot be modified once it is begun.

In these situations, we must go beyond the factors traditionally considered by economics and beyond any additional psychological characteristics that might be ascribed to economic man. Sociological principles like those used here must be considered if we are to effectively predict decisions made under uncertainty. Thus the conclusion about action under uncertainty has two parts. The first is simply that many human decisions are best characterized in the terms used in this study. The second is that, when the relevance of rationality is obviated by ignorance, fundamental considerations of rank and desire for rank may in large measure determine who among the blind will act in innovative ways.

Appendixes

The Fieldwork and the Data

THE DATA used in this study were, for the most part, gathered between September 1966 and June 1967. However, it would be grossly misleading to present the study as the result of nine months of fieldwork. Between 1960 and 1965 I spent more than twenty months doing fieldwork in Zinacantan, and from 1964 on my principal interest was in Zinacanteco corn farmers. (Earlier work is reported in Cancian 1965a.) Moreover, my work was done as part of Professor Evon Z. Vogt's Harvard Chiapas Project. Between 1958 and 1967 the project sponsored more than twenty man-years of fieldwork in Zinacantan (see Vogt 1969: Preface), and fieldworkers benefited from each other's work.

When my wife and I first went to Zinacantan in 1960 the number of experienced fieldworkers was small. Old-timers measured rapport in terms of the handful of cases in which an anthropologist had been invited into a Zinacanteco home; and only a few Zinacantecos had worked as paid informants. As will become apparent below, rapport blossomed rapidly over the years. Both anthropologists and Zinacantecos have constantly traded on the basis of relations with others in each group, and these networks of access and obligation now include hundreds of people. Without this background of rapport, it would have been impossible to do the present study in Zinacantan.

The data gathered may be divided into three types: general background on Zinacanteco life, especially corn farming; structural summaries of inputs and outputs in the work of some farmers; and complete surveys of two hamlets within the township of Zinacantan. The techniques used to gather the data varied from participant observation, through informal interviewing, to formal questionnaires administered by trained assistants. Under these conditions, the role usually described as the informant role must be broken down into several parts. I talked informally with hundreds of people, both Zinacantecos and Ladinos. In addition, I formally interviewed some thirty Zinacantecos who were paid for their time; these people may be called "informants." Finally, five Zinacantecos ("interviewers") were hired to administer questionnaires to more than 300 other Zinacantecos ("respondents").

General Background

Much of the general background for this study was gathered in the course of living in contact with Zinacantecos over a long period. Casual conversation can almost always go in the direction of shoptalk in Zinacantan, and this means that one is talking about corn prices, the availability of new land for rental, rain, and the progress of the current year's crop. I took a number of trips to the lowland areas where Zinacantecos farm, during which I saw work in the fields, negotiations with landowners, conversations with merchants, and the operation of the government receiving centers. I also chatted with landowners and other lowland people, walked and measured fields, checked maps, and confirmed information gathered in the highlands about life in the lowlands.

Structured Interview Information

Although successful corn farming depends on a good sense of cost accounting, correct judgments and decisions do not require that all the relevant factors be made explicit. In my interviews, there was always a tension between the informants' inclination and ability to make things explicit and my desire for systematic information about inputs and outputs. Occasionally, informants initiated patent lies; and occasionally, I lost control of myself and pressed for statements that overtaxed the informant's memory or knowledge. Obviously incorrect information resulting from these situations was thrown out before analysis began, and I have confidence in the gross accuracy of the information that is presented in this study.

The information from interviews was recorded in two structured, quantitative forms as well as in narrative. The first of these forms is the work history. The following information was recorded for as many years as each informant could remember: location of fields, original rent negotiated with the landowner, condition of the land relative to the fallow cycle, amount and type of seed used, total harvest and/or yield, amount paid as rent, disposal of the remaining corn, transport cost for men and corn to and from the fields, and wages paid to workers. Work histories were made out for 22 informants; two were obvious failures and were discarded, three were incomplete but useable, and the other seventeen were successful beyond my expectations. Fifteen of the twenty useable ones involved men who had farmed for more than ten years as independent adults. For the most part, however, only information on the period 1957–66 was used in my tabulations. More details on the use of these data are given in footnotes accompanying their analysis in the text.

The second form of structured reporting is the "work budget." These attempt to give fully detailed descriptions of inputs and outputs for the 1966 crop year or any other year the informant felt he could accurately reconstruct. This is a much more difficult task, and most informants were not able to complete it. Three of the most notable successes are reported in Appendix B. These data and other partial successes are used extensively in Chapter 4.

The Surveys

A survey was designed in order to get uniform and comparable information on all the adult males in two hamlets: Apas and Nachig. Information that is identified as survey information comes from three sources: interviews done by Zinacanteco interviewers using questionnaires; fairly complete genealogical information on both hamlets gathered from informants by myself and other anthropologists; and ratings by a limited number of informants of all the men included in the survey. Since the population studied was defined principally by the genealogical work, I will describe that first.

Since 1962, when I first did a crude census of Apas (see Cancian 1965a), George Collier and I have maintained an improved census and genealogical information on the hamlet. This work was facilitated by Collier's development of computer programs for processing genealogical materials (Collier 1969). The data were gathered from a limited number of informants who themselves informally interviewed other Zinacantecos when they were uncertain of details. The genealogical material was checked and revised in 1966 before the questionnaire part of the survey was made, and changed little during the questionnaire survey.

The Nachig census was begun during summer of 1966 by an informant under the direction of Collier, and was improved when I added genealogical information to it before the questionnaire was applied. The less refined nature of these data, the larger size and complexity of Nachig, and the difference in skill between Apas and Nachig informants is reflected in the fact that more change in the Nachig materials was necessary during the questionnaire survey. For example when the interviewers were instructed to recruit as a respondent any married male living in the hamlet of Nachig, they produced a number of people not previously included in our census.

Of the 147 men known to be living in Apas, 142 (97 percent) were interviewed. Some information on the other five was provided by informants, but data involving opinions or tests of an individual's knowledge were of course not available without an interview. Of the 245 men initially listed for Nachig, 232 (95 percent) were interviewed. Three others directly refused to be interviewed, and ten were never approached. If the nine additional men discovered in the genealogical work done after the questionnaire survey are added to the 245, the percentage interviewed drops to 91.

The questionnaire was principally designed, translated, and pretested under the supervision of my wife, Francesca, who was doing an independent study. These aspects of the work will be described in more detail in her forthcoming report (Cancian n.d.). The items that I added to the questionnaire are reproduced at the end of this appendix. We shared the work of organization during February–March 1967, when the questionnaire was applied.

The five native interviewers recruited respondents and brought them to a house near the San Cristóbal market that we had rented and equipped for the work. An interview usually lasted between thirty minutes and an hour; and respondents often waited in the courtyard when all the interviewers were busy. Respondents were paid eight pesos plus travel expenses, and also received a small gift. The selection of a gift from the eight items offered was part of the information required for my wife's study; but the fact that a gift was offered apparently added to the attractiveness of the "work." We did not expect to get anything approaching the virtually complete coverage of the hamlets that was finally achieved. Though we did lay the groundwork carefully so that the survey would not be hampered by political factionalism in the hamlets, credit for our success in recruiting respondents must go to the general rapport achieved by the anthropologists connected with the Harvard Chiapas Project, and especially to the skill and hard work of the interviewers we employed.

Since direct questions about wealth could not be asked on the questionnaire without upsetting the respondents and eliciting lies, this kind of information was provided by the interviewers from their personal knowledge of the respondents and from informal interviewing they did. Two measures of wealth were recorded. The first was size of cornfields planted, both in the highlands near home and in the lowlands where land is rented from non-Zinacantecos. The second was a relative wealth ranking generated by the interviewers in a card-sorting task. Only the first measure is used in this study, and only lowland fields were considered (except in Chapter 2).

Since young Zinacanteco men often work closely with their fathers for a short time after marriage, it was necessary to devise a way to eliminate their cases from the survey population before analysis. For the synchronic analysis (the year 1966) all men aged 24 or less, as well as all men over 24 who explicitly stated they still worked with their fathers, were excluded. For Apas this meant excluding 26 married men aged 24 or less and two married men over 24; for Nachig, 24 married men aged 24 or less were excluded. Thus the final survey populations with which the synchronic analysis began were 119 in Apas and 208 in Nachig. For the diachronic analysis (the period 1957–66), similar rules were used to eliminate men who would not have been 25 in the year in question and (in a handful of cases) men who were old enough but stated that they were still working with their fathers despite their age. The size of the final survey populations for each of the ten years was as follows:

Year	Nachig	Apas		Year	Nachig	Apas
1957	131	74		1962	171	104
1958	138	80		1963	186	111
1959	144	83		1964	190	113
1960	149	87		1965	196	115
1961	152	100		1966	208	119

With the exception of a very small number of "no information" cases for some variables, other variations in population size that appear in the tables and text are due to the elimination of cases that are not relevant to the question being asked. The most common such adjustment is the elimination of those who had no lowland fields in the year in question. In 1966, for example, this reduces the Apas population to 108 (11 with no fields) and the Nachig population to 193 (15 with no fields).

QUESTIONNAIRE ITEMS ON CORN FARMING

1. Do you grow corn?

2. Where did you farm this past year? How many years have you farmed at [name of ranch]? Did you farm elsewhere while you were farming there? Where did you farm in previous years? [names of ranches back to 1957]

3. Of all the places you have farmed, which was the best?

4. Have you heard of any better place? Have you been there? Who told you about it?

5. Who was the leader of your farming group(s) this past year? [name and hamlet]

6. How many of you were represented by him?

7. Where are you farming this year [1967]?

8. Who is the leader of your farming group(s) this year?

9. How many years have you been working apart from your father?

10. Do you hire men to work on your cornfield?

11. Do you work on other farmers' cornfields?

12. Where did you sell your corn last year? Two, three, four, and five years ago? [Categories were: direct sale in San Cristóbal; sale to dealer in San Cristóbal; sale to government center; sale to private buyer in the lowlands; sale to Zinacanteco speculator.]

13. Now we will talk about where the government receiving centers are. Is there one in Acala? Chiapa de Corzo? Soyatitán? Flores Magón? Venustiano Carranza? Colonia 20 de Noviembre? Chiapilla? [There were in fact no centers in Soyatitán and Chiapilla. If the answer was yes, the informant was asked if he had actually seen the center in the designated location.]

Three Zinacanteco Farmers

EVERY ZINACANTECO farmer presents a unique case; but whatever the peculiarities of his personal situation and the piece of land he farms, he is subject to the constraints described in this study. In the text above, universal tendencies are stressed. In this appendix, the farming activities of three individuals are described in detail for the 1966 crop year, and their uniqueness is allowed free reign. Although these cases cannot be said to be typical, the separate factors that characterize each are frequently important for other Zinacanteco farmers.

A complete description of the kind presented here was attempted for almost every one of the twenty informants from whom work histories were taken. Many of the attempts failed because the interview developed poorly, or because the informant was unwilling or unable to give complete information. One or two other cases besides the three chosen offer enough detail to be presented in this form; but most are complete for only a part of the farming cycle. It is probably not an accident that all three informants described here have major economic commitments to nonfarming enterprises that depend heavily on interpersonal relationships. In addition, two of them, Marian and Romin, belong to families with whom anthropologists have had contact for many years. Though they do not live in Nachig or Apas, their cases are included because of the detail they were able to give. The third informant is Antun, a Nachig man who had been an acquaintance for almost two years at the time of the interview. With one exception (noted below), the picture he gave was supported by internal checks and is consistent enough with standard practice to be taken as reliable.

These cases illustrate the whole range of factors relevant to a Zinacanteco farmer's economic situation, and they will hopefully give the reader a feeling for how the abstractions developed in the text may apply to an individual farmer.

Marian

Marian, a man in his late forties, comes from a family of salt traders in the Zinacantan ceremonial center, Hteklum.[1] As a youth he followed his father in the salt trade, taking over the territory that by tradition belonged to his family. In the early 1950's there was a near-famine period during which corn prices skyrocketed, and Marian decided to insure his livelihood by farming as well as selling salt. He began in 1953 by seeding one almud at a ranch called Trapichito in Zone 2. By 1956 he was seeding two almuds, and in 1961 he began seeding three almuds. The year reported here, 1966, was the first in which he worked at more than one location and the first in which he seeded more than three almuds.

In 1966, Marian seeded two almuds at Rodeo, a ranch very close to the Zone 1 secondary road from Chiapa de Corzo to Acala, and reduced his seeding at Trapichito to two almuds. He began at Rodeo, he says, because he was told it offered good land, and because it was near the road, so that his corn could be easily transported. The crop at Rodeo was poor, even by Zone 1 standards, and in 1967 Marian again concentrated all his effort at Trapichito. Tables B.1 and B.2 give detailed figures on inputs and yields for 1966, and should be read with the discussion that follows.

Arrangements for renting the land at Trapichito are made directly with the owner, a widow who lives in San Cristóbal. During his fourteen years at Trapichito Marian farmed six different plots, each for two or three years. Each plot had been used by others recently, usually in the year before he began to farm it. To change plots, he simply got the owner's permission in San Cristóbal and then talked with her representative at the ranch to determine exactly what land he would have. Trapichito's rental pattern, which is unique and seems to be a holdover from the period before land reform, includes work on the owner's personal fields as well as rent payments in cash and corn. In 1966 the rent per almud seeded at Trapichito was: (1) $40 in cash paid in advance in January; (2) six almuds of corn delivered to the owner's residence in San Cristóbal after the harvest; (3) the following days of work in the owner's fields.

Preparing the land: 0.5   Doubling: 0.5
Seeding: 0.5   Gathering (harvest): 1.0
First weeding: 1.0   Threshing (harvest): 0.5
Second weeding: 1.0

[1] Marian is the man who took the A7S religious cargo in 1961 (Cancian 1965a: 81, Table 5). He is Domingo's father (Cancian 1965a: 121–22).

TABLE B.1
Marian's Work Budget for Trapichito Ranch
(two almuds seeded)

Category	Prepare land	Seed	Reseed	First weeding^a	Second weeding	Double	Harvest
Number of workers	—	1	—	3	3	1	3
Man-days worked:							
Farmer	3.0	2.5	1.0	4.0	3.0	1.5	15.0
Workers	—	2.5	—	12.0	9.0	1.5	42.0
Farmer (for owner)	1.0	0.5	—	0.5	0.5	0.5	—
Workers (for owner)	—	0.5	—	1.5	1.5	0.5	3.0
Expenses (pesos/almuds):							
Pay per week	—	$30/	—	$30/4	$30/4	$30/	/4
Total pay	—	15/	—	30/4	30/4	10/	/30
Transport of pay^b	—	—	—	—	—	—	—
Truck fare (farmer)	—	—	—	—	—	—	—
Truck fare (workers)	—	—	—	—	—	—	—

NOTE: A dash (—) in a cell means none or not relevant. Other data of interest are as follows. *Travel time:* two days per man-trip. Marian made the seven trips shown in the table plus one to inspect his fields in August. Workers made eleven man-trips. *Harvest:* 240 almuds plus four almuds of beans. *Seed:* Two almuds plus one almud of beans. *Rent:* $80 plus 12 almuds plus the work shown. *Transport of harvest:* $276. Marian sold ten almuds in the lowlands, got beans on a rented truck free, and got about ten almuds in each bag of corn at $12 shipping cost per bag. His rent in corn was taken to San Cristóbal.

  a Other expenses for this weeding were: 2.67 almuds of corn and .33 almud of beans to feed workers; $3 to recruit workers; and $15 for the customary meal after returning from the lowlands.

  b For the harvest, he delivered pay to two Chamula workers with his horse, and it took less than two full days. Workers hired at other times were Zinacantecos from Hteklum.

Everything calculated, this complicated arrangement is less expensive than the 24 almuds per almud seeded paid by most Zinacantecos, but it also reduces a farmer's free time.

At Rodeo, Marian worked with a small group, and in 1966 the group leader's yield was even lower than Marian's. (In 1967 the leader abandoned lowland farming and concentrated on growing flowers for the San Cristóbal and Tuxtla markets.) The rent at Rodeo was paid in kind, as is usual. The rate was 18 almuds per tablon, which is low; and the land was measured generously, so that Marian was able to seed two almuds on the 1.5 "tablons" for which he contracted. Thus he agreed to pay 27 almuds for all the land he farmed, an effective rate of 13.5 almuds per almud seeded. The crop was so bad that the owner agreed to reduce the rent from 27 to 20 almuds, which made it a little less than 25 percent of the crop.

Marian used workers more than one would expect, given the size of

TABLE B.2
Marian's Work Budget for Rodeo
(two almuds seeded)

Category	Prepare[a] and seed	Reseed	First weed-ing	Second weed-ing	Double	Har-vest
Number of workers	1	—	3	3	1	2
Man-days worked:						
Farmer	4	1	3	3	1.5	7
Workers	4	—	9	9	1.5	14
Expenses (pesos/almuds):						
Pay per week	$30/	—	/4	/4	$30/	/4.00
Total pay	20/	—	/6	/6	7.5/	/9.33
Transport of pay	—	—	—	—	—	—
Truck fare (farmer)	7/[b]	6/	6/	6/	6.0/	3/[c]
Truck fare (workers)	6/	—	18/	18/	6.0/	6/[c]

NOTE: Other data are as follows. *Travel time:* Two days per man-trip for four steps: preparation and seeding, first weeding, second weeding, and harvest. Marian spent one day traveling for the reseeding and one day traveling to look at the fields in August; he and his workers spent 1.5 days each in travel for the doubling. *Travel cost:* Marian spent $6 truck fare to look at the fields. *Harvest:* 90 almuds. *Seed:* 2 almuds. *Rent:* 20 almuds. *Transport of harvest:* $49 ($7 per ten-almud bag). I have no information on food, recruitment, and meal-on-return expenses.
[a] Preparing land and seeding done on same trip.
[b] Marian paid an extra peso for the seed corn he was carrying.
[c] They all rode back free with the corn.

his operation. For instance, he personally worked only 14 days on weeding and 3.5 days on doubling. In part, these figures may reflect the inefficiencies of traveling between two widely separated locations and the demands of his salt business. However, a number of personal factors may also be relevant, especially in his use of workers at Trapichito. Since Marian is not part of a work group at Trapichito, he may take workers simply for company in the lowlands. For the seeding and weeding at Trapichito he employed his impoverished younger brother, who is a near-mute and has no lowland fields. For the weeding he hired in addition a ritual kinsman from Hteklum and the kinsman's son. Neither had fields in the lowlands, and I suspect one would not have come without the other. Marian pays the lowest possible wages and offers few extras in food and drink; hence he probably has to choose workers from a relatively small pool. Moreover, he is a quiet man, and is apparently not given to the dynamic organizing required of a big operator who takes many relatively anonymous workers to the lowlands.

At Trapichito, Marian's work pattern was the standard one, with eight trips to the lowlands (including one for reseeding and one for looking at the fields in August). His labor input for preparing the land, seeding, weeding, and doubling was a little below the standard figures. The harvest of the 240-almud crop would have taken thirty man-days with standard inputs, and the harvest of his small bean crop (see below) took six days by his own calculations. Thus the 42 man-day total for the harvest is about 17 percent more than standard figures.

At Rodeo, Marian's work pattern was also the standard one, with one exception that is frequent on old land: he cleared the land, burned it off, and seeded during the same trip. His labor input for weeding was low. (Looking at the land in the Rodeo area, however, one gets the impression that nothing, not even weeds, grows in abundance.) The labor input he reported for the harvest is extremely high for the amount harvested (21 man-days for 7.5 fanegas); on the other hand, it is slightly lower than that for a normal crop on the same amount of land. Thus, although part of the difference between Marian's statements and the norms for harvesting may be due to a reporting error, a large part of it is certainly due to the inefficiencies of harvesting a meagre crop.

Since Trapichito is relatively high up above the valley floor, Marian used the highland technique of seeding beans and corn together (two beans and about six grains of corn in the same hole). His yield of beans in 1966 (four almuds for the one-third almud he seeded) was relatively low, he said. But in any case, the cost of harvesting this type of bean, as he reports it, is so high (1.5 man-days for an almud with a market value of $15) that profits are very low. The extra labor input for beans planted with the corn, however, is nominal, so planting them brings a small but relatively sure profit.

Marian and his workers always walked to Trapichito, taking a full day each way; and he took his horse as a pack animal on every trip but the ones for reseeding and looking at the fields. In traveling to Rodeo, he walked to the Pan-American Highway in Nachig and picked up a ride to the crossroads near Chiapa de Corzo on one of the many passing vehicles. The fare for this ride is relatively low ($3), and the walk would be very long. Going to Trapichito by vehicle would involve the same initial walk, a long and expensive ride past Acala on the secondary road, and a walk back up the mountains to his fields.

For transporting his harvest, however, Marian used trucks in both locations. The cost from Rodeo to Hteklum is $7 per sack (9–10 almuds). Had the harvests been comparable, this would have represented a great advantage over transport from Trapichito, which costs $12 per sack. Although the cost of motor transport from Trapichito comes to about $1.20 per almud, Marian has found that in the end this is less expensive and bothersome than hiring mule transport from other Zinacantecos, who charge $1.10 per almud. Mules must be fed corn by the

farmer, their owner must be fed well at the end of each trip, and the
farmer must spend two days making each round trip. Since each mule
carries only eight almuds per trip, even a relatively large string of ten
mules carries only eighty almuds; and the crops that Marian normally
gets would thus require at least two trips. By using a truck, he saves
time, corn, and meals for the mule owner; and he gets a free ride home
with his crop.

On the whole, Marian's farming operation is a successful one, especially if we ignore his brief experiment at Rodeo. He is what I have
called a "holding operator" (Cancian 1965b): his approach is one of
small cautious risks and carefully considered small investments made
to achieve small gains. He is a respected man of moderate means who
has taken a respectable but inexpensive religious cargo in the community. Since he may never take a second cargo, his needs are not
great. His one son by his first wife is well established with a family of
his own, and Marian lives alone with his second wife, who has given
him no children. He has a few relatives and in-laws living nearby, but
none depend on him economically and none are large scale corn farmers.[2] In sum, Marian is a solid citizen, quietly settled in a respected position in the community. His corn farming is an important part of maintaining this status.

Romin

Romin's farming patterns are greatly influenced by his membership
in the Vaskis lineage (see Chapter 3). In December 1966, when he
was 26 years old, he became the first male to marry into the third generation of the lineage. As we have seen (Table 3.4), he had been involved with the Vaskis farming group for about ten years before his
marriage. To be exact, he said in answer to a direct question, he got
the land first and the woman second. In 1960 he began farming with old
Shun's group, and later the same year he became engaged to Shun's
granddaughter. The price of admission to this powerful lineage was
high, and his courtship was a long one.[3]

Romin began his farming career working with his father at a location

[2] One of his three brothers-in-law is an occasional corn farmer and trader,
and has worked on coastal coffee plantations (an unusual occupation for a
Zinacanteco). Another is a dynamic, controversial man who is a full-time salt
trader. The third brother-in-law is a holding operator and trader who has been
more successful economically than Marian. This man, another Marian, was
one of the five principal informants in my earlier study of the Zinacanteco
cargo system (Cancian 1965a).

[3] See J. Collier 1968 for a study of courtship and marriage in Zinacantan.

in the distant part of Zone 3. After his father died in 1956, Romin left that location and farmed in several places (as described in Chapter 3) before settling down with the Vaskis group. Romin's father had been a rich man, and before his untimely death he had held important and expensive religious posts in Hteklum (cargos A2J and B1, see Cancian 1965a). Romin himself has not yet established a position in the community. Most of his work before 1966 was devoted to preparing for the enormous expenses of his wedding to a girl from a wealthy and well-known family. During the last two or three years before his wedding, his income from farming and a bit of trading was supplemented by substantial amounts he earned working for anthropologists connected with the Harvard Chiapas Project. His value as an informant was greatly enhanced by his ability to write. In this way, and in many others that will no doubt develop in the future, he provides the lineage with an additional source of strength.

The leader of the Vaskis work group is old Shun, and the group's movements are described in Chapter 3. Rent at Río Trapich, where Romin farmed with the group in 1966, is like that paid at Rodeo by Marian. Romin contracted for 1.5 tablons of land at 18 almuds per tablon in that year, and managed to seed two almuds on the area assigned him. Since his crop was close to normal, he paid the full 27 almuds of rent (see Table B.3).

Romin's use of workers was normal in that he used them only at weeding and harvest. It was very abnormal in that he did not work himself during the second weeding and harvest; instead, he worked for anthropologists, earning more than he could by working in his fields during these periods. Since his fields were adjacent to those of his future relatives, who wanted him to be well-prepared financially for his wedding, he was able to leave the supervision of his workers in their hands. His marriage plans also gave him advantages in the recruitment of workers for weeding, since his workers were hired from a large group of Chamulas who came to work for various members of the Vaskis group. Although I have no substantial evidence on this point, it seems that the reputation and reliability of the Vaskis lineage attracts a steady supply of hired labor.

The labor inputs Romin reported for his fields are close to normal with the exception of the harvest, where the input was 16 man-days for a crop that would take 20.25 man-days with standard inputs. Whether this is a reporting error or a reflection of the fact that Romin paid a premium wage of six almuds per week I cannot tell.

Since his land is less than fifty yards from the Chiapa-Acala road and his house even closer to the Pan-American Highway in Navenchauc,

TABLE B.3
Romin's Work Budget for Río Trapich
(two almuds seeded)

Category	Pre- pare land	Seed	Re- seed	First weed- ing	Second weed- ing	Double	Har- vest
Number of workers	—	—	—	2	2	—	2
Man-days worked:							
Farmer	8	6	3	12	—	6	—
Workers	—	—	—	24	18	—	16
Expenses (pesos/almuds):							
Pay per week	—	—	—	/5	/5	—	/6
Total pay	—	—	—	/20	/15	—	/16
Transport of pay	—	—	—	$12/	$8/	—	—
Truck fare (farmer)	—	$10/	$10/	10/	—	$10/	$5.0/[a]
Truck fare (workers)	—	—	—	8/[b]	8/[b]	—	—
Food, corn	/.83	/.67	/.44	/4.33	/2.67	/.67	/2.17
Food, beans	—	—	—	/.67	/.33	—	/.33
Food, extras	$10/	10/	5/	—	—	10/	10/
Recruitment	—	—	—	2.5/	2.5/	—	—
Return meal	—	—	—	16/	16/	—	12.5/

NOTE: Other data are as follows. *Travel time*: Two days per man-trip, plus two days Romin spent visiting at harvesttime, plus one day Romin spent looking at the fields in August, plus two man-days for the worker listed below under "other work." *Travel cost*: Romin spent $10 traveling to look at his fields. *Other work*: Late in the season Romin hired his sister's husband to spend three days mending fences; this cost him $21 cash, .44 almud of corn, and a small quantity of beans. He had no expenses for recruiting or extras. *Harvest*: 162 almuds. *Seed*: 2 almuds. *Rent*: 27 almuds. *Transport of harvest*: To Navenchauc, $75 total.
  [a] Estimated by F. C.
  [b] From San Cristóbal to Navenchauc before walking to lowlands; and return to San Cristóbal after work was done.

Romin did not have to make special arrangements to transport his corn home. He simply hailed a passing truck, which took his 15 sacks of grain to the crossroads near Chiapa for $30; there he waited for another truck, which took them on to Navenchauc for another $45.

Given his heavy schedule of work for anthropologists, Romin's own labor contribution was not a normal one. Moreover, his investments in personal truck fare and extra food reflect the fact that he had this source of ready cash. This unusual situation only serves to highlight the advantages of belonging to a farming group made up of kinsmen. Though most men would not normally draw on kin ties to the extent Romin did in 1966, having them can be important insurance for the Zinacanteco farmer.

Anticipating a hiatus in anthropological work during late 1967, Romin used some of his remaining cash and some loans to buy land in the highlands, on which he planned to grow potatoes and flowers for trading. In addition, he planned to seed beans with his corn in the fall of 1967. In the long run he may take up the pattern that seems to characterize the Vaskis lineage. Given their apparent wealth and their position in the community, the Vaskis men seed relatively little corn in the lowlands (about three almuds each); but they cultivate flowers, fruit, vegetables, and corn on their highland and ejido land, and they occupy themselves in trading when there is no work in the cornfields. The economic potential of this pattern is reflected in their exceptionally active —and expensive—participation in the system of religious offices (Cancian 1965a: 120).

Antun

Antun's principal occupation is retailing corn in the San Cristóbal market (he is one of the five men in the core group of retailers mentioned in Chapter 5). Despite this extensive nonagricultural activity, his farming operation was the most successful of the three described here.[4]

Antun is 28, lives in Nachig, and was married in December 1959. In 1962 he went to the Merced Ranch (Zone 5) as a worker for another Nachig man. He saw that land was available, he says, and asked for some to seed the next year. At first only one Nachig man joined his "group," but before the 1963 season started one other Nachig man and two brothers from Chamula also asked him to get land for them. The group's membership remained the same for the four years it farmed at Merced. The rent was the standard 24 almuds per tablon, and each tablon apparently took the normal almud of seed.

Except for the bean harvest, when he recruited his group members to work for him, Antun did not use Zinacanteco workers. He paid the going wage of five almuds per week, and was generous in paying the truck fare of workers from his house to theirs (not a standard expense) and in providing extra food (mostly meat) in the lowlands. His recruitment of workers is reviewed in detail below.

Like Marian at Rodeo, Antun prepared his land and seeded during the same trip. The time he spent on these steps of the agricultural cycle is quite high, and I can only attribute this to the fact that the land he farmed had been planted in 1963 (by him) but had been used as pasture in 1964 and 1965. Antun's work pattern differs from the standard

---

[4] A contradiction in my notes suggests that Antun's harvest may have been less than the very good 150 almuds per almud seeded shown in the accompanying tables. However, even if the lower figure is used he is still the most successful farmer.

one at two points. (1) As group leader, he made a trip in February to make certain that his group could have the land it had anticipated getting at the end of the previous season. He travelled by truck on this occasion, he said, because he does not like to walk alone in the lowlands; and during the trip he put in two days of work repairing fences. (2) Since he began seeding beans on September 1, he did not take a

TABLE B.4

Antun's Work Budget for Merced

(three almuds seeded)

Category	Prepare[a] and seed	First weed- ing	Second weed- ing	Seed[b] beans	Harvest	Harvest[c] beans
Number of workers	2	3	3	3	5	3
Man-days worked:						
Farmer	13	10	6	10	9	4
Workers	26	30	18	30	45	12
Expenses (pesos/almuds)						
Pay per week	$36/5	$36/5	/5	$36/	$36/	$36/
Total pay	12/20	72/15	/15	180/	270/	72/
Transport of pay	13.20/	12/	$12/	—	—	—
Truck fare (farmer)	—	—	—	—	—	—
Truck fare (workers)[d]	—	15/	—	18/	—	—
Food, corn	/5	/6.33	/3.67	—	—	—
Food, beans[e]	/1	/2	—	—	—	—
Food, extras	45/	40/	10/	—	—	—
Recruitment	4/	7/	8.50/	4/	8/	—
Return meal	20[f]	31.50/	16.70/	19/	25/[g]	—

NOTE: Other data are as follows. *Travel time*: Two days per man-trip for all steps before harvest, plus one day travel for Antun to secure land. See detailed schedule of harvest period for travel. *Travel cost*: Antun spent $18 for truck fare when he went to secure the land. *Other work*: Antun worked two days mending fences when he went to secure the land. *Harvest*: 448 almuds plus 48 almuds of beans. *Seed*: 3 almuds plus 3 almuds of beans. *Rent*: 72 almuds plus 3 almuds of beans. *Transport of harvest*: Corn transport cost $150 for 30 bags sold to the receiving center, nothing for 24 almuds of corn he sold to private buyers, and $100 for 88 almuds he took home. He paid $40 to transport himself and 36 almuds of beans to Chiapa de Corzo, and $10 for the nine almuds he took home. *Gross income from sale*: Corn sold to the receiving center brought $2,935 for 30 bags, or 264 almuds (he measured the standard bag at 8.8 almuds). Corn sold privately brought $190. And he received $809 ($1.70 per kilo) for the beans he sold in Chiapa.

[a] Preparing land and seeding done on same trip. Antun did not have to reseed.

[b] Antun doubled his corn when he went to seed beans in early September. Of the ten days (40 man-days) shown in this column, three (12 man-days) were devoted to doubling.

[c] The harvesting of beans may be separated from the harvesting of corn. See the detailed schedule of harvest activities.

[d] Truck fare he gave workers to get from his home to theirs after returning from the lowlands.

[e] He used the beans left over from the prepare/seed trip and the first weeding trip to feed workers on the second weeding trip.

[f] Figuring a conservative $15 for the chicken he took from his flock in place of buying beef for the meal.

[g] In the form of a cash bonus given in the lowlands.

trip to see his fields in late August. Moreover, he did the doubling himself during the same trip.

In Table B.4 I have listed labor input for the harvest as the work Antun and his workers did December 8–17. But as can be seen in the following description of Antun's activity during the harvest period, he worked several extra days that might also be counted as part of the harvest input (picking up loose corn, degraining seed corn, delivering rent, and sacking corn). Because he had a large crop and worked alone on these tasks, they show up as full days in this detailed account. A man with a smaller crop and workers helping him could have accomplished any or all of them in part of a day.

As shown in Table B.4, Antun's bean crop was an immense success financially. Normally, the beans would have been weeded shortly after planting. But Antun was commited to duties as a member of the Nachig School Committee on Mexico's Independence Day (September 16); he missed the ideal time for weeding beans, and decided not to do it at all. If this reduced his yield of beans, it did not do so enough to destroy his profits. The almud rent per almud of beans seeded that he paid is common on lands that are regarded as especially good for beans. In other places landowners allow beans to be seeded without additional rent being paid.

Since he usually walked to his fields with his workers, and since he sold most of his crop at the government warehouse in Acala, Antun's expenses for transport were relatively low.

Antun's status as a corn dealer in the San Cristóbal market adds significantly to his income. It also greatly complicates his relations with other Zinacantecos and makes it impossible to call him a typical corn farmer. However, the manner in which he organizes his corn farming operation can be seen as fairly typical of the larger operators. The extensive use of cash described in the section that follows is no doubt attributable to the size of Antun's operation and to the fact that as a dealer he is accustomed to handling large amounts of cash.

Antun's Activities at Harvest Time

A day-by-day account of Antun's activities from December 6, 1966 (Day 1), through January 10, 1967 (Day 36), when I interviewed him, illustrates many of the farming activities that cannot be easily classified. It also illustrates the concrete details of sale to the government receiving centers.

Day 1. Five workers come to sleep at Antun's house.
Day 2. All six travel to the lowlands by foot.
Days 3–8. They work gathering ears of corn. On Day 6, a Sunday,

Antun rests; but the workers continue, so that they will not lose a day of pay.

Days 9–11. All six thresh corn. After this, the principal work is done, and the workers want to look for other jobs in the lowlands. They are given their pay in cash and a bonus of $5 each in lieu of a meal on return to the highlands. Antun sells 24 almuds from the harvest for $190 and adds $105 he has with him to provide their pay.

Day 12. Antun finishes threshing.

Days 13–15. Antun cleans the corn (blowing out dust).

Day 16. Antun goes to a bank in Tuxtla and deposits $180 for thirty official sacks to package corn for a government receiving center. He picks up the sacks at the Acala center on his way back to his fields. Total travel expense, $14.

Day 17. Antun spends the day sacking his corn.

Day 18. He finishes filling the sacks about 2 P.M. and goes to Acala to hire a truck to take them to the center. Travel cost $2. He returns to the fields with the truck about 6 P.M., and his group members help him load it (supposedly in return for similar help later). Antun returns to Acala with the truck and sleeps on his corn to guard it; the driver returns to his nearby Acala home on foot.

Day 19. Antun is first in line at the warehouse and finishes his business by 8:30 A.M. The cashier tells Antun to return the day after tomorrow for his money (the next day is Sunday and Christmas). Antun spends the day in Acala, and buys meat, shirts, and string to sew up sacks for members of his group who have requested them. He returns to his fields about 6 P.M. Travel cost $2.

Day 20. Antun gathers the loose grain left from degraining.

Day 21. He goes to the center and spends the day waiting to be paid. He is finally paid at 6 P.M., when he insists that he is from the highlands (i.e., far away) and needs the money to eat. Local people wait even longer. He returns to his fields late, after stopping to pay the truck driver who took his corn to the center. Travel cost $4.

Days 22–23. He gathers up loose grain and degrains the corn he has selected for seed.

Day 24. He spends the day carrying his rent to the landowner's ranch house with his horse. His group members are not yet ready to pay their rent.

Day 25. He finishes sacking his remaining corn and leaves it stacked.

Day 26. He goes home to Nachig by truck. Travel cost $9.

Day 27. He goes to the fiesta in Hteklum, the ceremonial center, to visit a brother-in-law who has entered a religious cargo.

Day 28. He returns to his fields by truck and recruits members of his group to help him harvest his beans. Travel cost $9.

Days 29–32. Antun and three workers harvest beans. On the last day he pays the workers in cash.

Day 33. Antun goes to Chiapa de Corzo to sell the beans. He borrows horses to get them to the road and hails a passing truck. Having sold them, he returns to his fields, picks up the remaining corn and beans, and goes home to Nachig by truck, arriving at 6 A.M. the next day.

Day 34. In the afternoon he goes with his brother to his brother's fields in Zone 9. They go with the truck his brother has hired to bring up his corn.

Day 35. They return by truck.

Day 36. Antun spends the day working with Frank Cancian.

Antun's Activity Recruiting Workers

Antun runs a relatively large farming operation, and he has a relatively small number of relatives in Zinacantan. Being young, and being a dealer, he has few ritual kinship or friendship ties that would enable him to recruit workers in a particularistic way. For all these reasons, he hires workers from the highland communities near Zinacantan on a universalistic basis. As can be seen in Tables 3.9 and B.1 to B.4, his expenditures for extras for these workers is high compared to those of Marian and Romin, who hire workers in more particularistic ways.

The following description of the way Antun hired workers for each of the steps in the agricultural cycle illustrates the operation of a relatively large-scale farming enterprise, as well as the relation of Zinacanteco farmers to workers from other communities.

Preparing the land and seeding. About ten days before making the trip in early May, Antun went to the houses of two Chamulas that he knew and recruited them to help with his work. He gave each a present of liquor. They agreed to come to sleep at his house the night before the trip to the lowlands. On returning from the fields they had a meal at his house and went on home. He later took their pay in corn to San Cristóbal and got a truck driver to deliver the corn to their homes in Chamula on a regular run that went that way. They returned the sacks to him later when they were passing his house in Nachig.

First weeding. He recruited workers about June 1 for the trip to the lowlands on June 10, and they agreed to come to his house on June 9. Since the workers who had helped him with preparation of the land and seeding already had commitments for this period, he hired three others. He found them by going to the point where one of the paths from Chamula enters the outskirts of San Cristóbal and hailing a group of three Chamulas who passed. They agreed to work, and he gave two of them who were brothers a liter of strong liquor. The third man, who did not drink, was given $2 in cash in lieu of a gift. On returning, the workers slept in Antun's house after the meal. The next day they went home, taking their pay in corn with them. Antun paid their truck fare and the cost of transporting their corn.

Second weeding. At the end of the first weeding trip, Antun recruited the same men for the work of the second weeding (about two weeks later). The two brothers who drank received a bottle of liquor each, and the nondrinker received a gift of sweet bread and soda pop. All received the usual meal after returning from the lowlands, and Antun paid to have their pay in corn taken to their homes. Apparently he did not pay the truck fare of the men for this trip; his generosity in paying this fare after the first weeding was probably part of his effort to recruit the workers for the second weeding.

Seeding beans. For the work of doubling the corn and seeding beans in early September, Antun hired three other men from Chamula who came to his house looking for work. He gave them a liter of liquor between them to seal the bargain, which cost him about $1.35 per worker (gifts to his previous workers had averaged about $2.40 per worker). At this time of year there is very little work, and it is easy to recruit workers. After returning from the lowlands they had a meal, and Antun gave them each pay in cash and truck fare home.

Harvesting. The Chamulas who had helped Antun seed beans had contracts to work on coffee plantations on the Pacific coast of Chiapas and were not available at harvest time, so he recruited five men from the more distant township of Chenalhó. The Chenalhó men did not have the official papers necessary to get coffee contracts. As noted in the description of Antun's schedule during harvest, when they finished their work for him they took a cash bonus in lieu of the final meal and went on immediately to seek other harvest work in the lowlands.

Weights and Measures

THE ALMUD, fifteen metric liters of volume, is used as the standard measure in this study. This use of a single unit simplifies the discussion and makes internal comparisons easy; but it does not permit comparison with standard measures of weight and area used in other places, and it fails to give a correct impression of the variety of measures used in the area where Zinacantecos live and farm. This appendix will try to compensate for both deficiencies, and will, hopefully, justify my choice of a single unit.

Zinacantecos are both farmers and merchants. As merchants, they are concerned with the accurate measurement of a relatively uniform commodity, corn. As farmers, they are concerned with getting the most out of the sloping, rocky land of uneven quality that they farm, and they cannot be as concerned with accurate measurement as a student of their agriculture might wish. The demands of both the merchant role and the farmer role are reflected in what follows.

Volume Measures of Corn

The following units are in most common use:

1 *caldera* = 1 metric liter
1 *cuarto* = 5 calderas (5 liters)
1 *almud* = 3 cuartos (15 liters)
1 *fanega* = 12 almuds (180 liters)

The *litro* (20 liters) and *fanega-litro* (12 litros, or 240 liters) are also used in many parts of the area. Around Acala the *cuartilla* (one-quarter of an almud) is sometimes used. When in the lowlands, Zinacantecos usually carry a general-purpose enamelware bowl called a *tasa*. As a corn measure it is equivalent to one-third of a cuarto.

The caldera measure is a metal cylinder usually purchased from metalsmiths in San Cristóbal (Photo 21). Cuarto, almud, and litro measures are typically wooden boxes made up by carpenters, although a five-liter oil can with the top removed is often used as a cuarto measure in the home. Two relatively standard sizes of burlap bag hold 6 almuds and 9–10 almuds respectively, but these are not used as measures in sale of corn. In the San Cristóbal market corn is typically priced by the cuarto, sold by the cuarto, almud, or litro, and measured out by the caldera. Zinacantecos almost always talk in terms of the largest applicable unit. Thus the rent paid per almud seeded in Zone 2 is expressed as 2 fanegas, not 24 almuds; similarly, the rent in Zone 9 is 2 fanegas-litro, not 32 almuds. The standard yields used to estimate profits in Chapter 6 are 2, 4, 6, 8, 10, 12, and 14 fanegas, in keeping with this local manner of expressing approximations.

Leaving aside the intentional use of inaccurate measures (see Chapter 5), the caldera and tasa measures are often rendered inaccurate as standards by the local custom of heaping them at the top, since their official capacity is based on a level measure. Capriata (1965: 8) reports that the Zinacanteco producers he bought from in San Cristóbal gave an average of 3.4 percent more than the official measure. A careful test I made of the use of the tasa measure suggests that it often yields almost 10 percent more than the official standard, though the informant who performed the experiment was able to reduce the "excess" of the tasa measurement to about 2 percent by heaping the official caldera against which it was being measured.

Since the difference between an almud of seed measured out with a tasa and that measured out with a level caldera is substantial, I have distinguished between an official almud and a farmer's almud in the following discussion of seed volume to land area equivalences. At harvest time the volume of the crop is apt to be measured with carefully leveled cuarto or almud boxes, if it is measured at all; but most farmers simply report the number of standard bags that they used to transport their harvest. The essential equivalence of informants' statements of how many fanegas make a ton of corn and my own estimates (based on weighing volumes measured with official calderas) suggests that the heaping custom applies mostly in the measurement of seed corn, and to a lesser degree in retail sales in the San Cristóbal market.

The Weight of Corn

Different types of corn weigh different amounts. Zinacanteco and Ladino farmers say that the Hybrid 503 used by some of them weighs a bit more than the flat, white, native grain (*pacha*) seeded by most Zinacantecos. The round yellow grain (*bola*) more typical of the high-

21. Left to right: tasa, cuarto, and caldera measures.

22. Zinacantecos use a stick the length of an armspan to measure off 25 armspans on a rope, which will in turn be used to measure a field.

23. Productive, flat land that produced an excellent corn and bean crop after being worked with plow technology.

24. Romin's land.

lands is universally recognized as weighing substantially more. The differences between the lowland varieties I weighed were small, and all tests indicated the essential accuracy of the common Ladino statement that there are 130 kilos to a fanega. This means that the weight of the official almud may be taken as 10.8 kilos.

Measures of Land

Land area is expressed in terms of the *tablon*, which is customarily 25 by 100 armspans (*brazadas*). The armspan, of course, varies with the man. A short landowner's measurement of armspans on a rope[1] averaged 1.6 meters per span, and my measurement of a tablon he had marked out on his property confirmed his use of this standard. A moderately tall Zinacanteco averaged 1.73 meters when trying to produce an honest measure and 1.78 meters when straining to make the measure long. A local agricultural engineer reported that in his experience the armspan averaged 1.72 meters. Two other Zinacantecos averaged 1.59 and 1.69 meters when trying to produce full, but honest, measures, and a second landowner said that the armspan should be 1.5 to 1.6 meters. A tablon measured with an armspan of 1.6 meters is .64 hectares; an armspan of 1.65 meters gives .68 hectares, and one of 1.7 meters gives .72 hectares.

Land Area and Volume of Seed

The size of a tablon may vary by as much as 15 percent, depending on whose armspan is accepted as the standard; but the land's capacity to receive a standard amount of seed varies even more than that, since it is usually sloping and rocky. Zinacantecos universally state that one almud seeds a tablon of land, and usually add that a tablon of very good land will take a litro of seed (1.33 almuds).

Table C.1 gives actual figures for three cases in which I am reasonably sure the measurements are accurate. The most impressive part of the figures is their variation; but when they are considered with other information they reveal an interesting pattern. Domingo was consciously given a tablon that was 25 by 119 armspans (though the table shows figures for one of 25 by 100), and he was able to seed 1.17 "farmer's almuds." The quantity seeded on the area given as a tablon thus decreases from Romin to Domingo to Juan. On the other hand the anticipated yield of the land at the time when the measurements were made

---

[1] A rope 25 armspans in length is customarily used to measure the plot. Sometimes the rope is measured carefully against a stick that has been cut to an armspan's length (see Photo 22). This practice presumably prevents lengthening the armspan by pulling the rope across the shoulders and throwing the arms back and down.

TABLE C.1

Land Area and Volume of Seed

Farmer	Hectares per official almud[a]	Hectares per actual tablon[b]	Farmers' almuds per tablon[c]
Romin	.52	.76	1.33
Domingo	.68	.73	.98
Juan	.76	.84	1.00

[a] Hectares of land required to plant 15 liters.
[b] Hectares of land in the area measured as one tablon.
[c] Amount of seed corn, by farmers' measure, that was planted in the area measured as one tablon.

by farmer and owner increases from Romin to Domingo to Juan. Romin's land is in Zone 2 near Zone 1, and he pays only 18 almuds of rent per tablon (see Photo 23). Domingo and Juan both farmed on the same large ranch in Zone 5 and paid the standard rent of 24 almuds. But Domingo's land had lain fallow only two years when he contracted for it (see Photo 10, p. 57), whereas Juan's land had to be cleared of mature forest (see Photo 9, p. 57). Juan's land produced well, though it almost seemed to be more limestone outcroppings than soil.

The levelest, least rocky land that I saw rented to a Zinacanteco at standard rent was an actual tablon of .64 hectares (see Photo 24). Thus, although the absolute area of a tablon tends to vary greatly, the variation seems to depend in part on the quality of the land. A farmer can depend on getting a farmer's almud into the land he is given as a tablon, and he may be able to seed substantially more.

APPENDIX D

Economic Man and Economic Change

THE QUESTION of whether Zinacantecos are economic maximizers or prisoners of traditions that inhibit their response to the economic opportunities resulting from government programs was hastily brushed aside in Chapter 1. There, I stated my belief that this is a bogus question that leads to scientifically incorrect and politically dangerous descriptions of peasant societies, and asserted my interest in the more limited uncertainty question, which required my viewing Zinacantecos as economic maximizers. Since my decision may seem unwarranted to many readers, and since the economic man versus traditional man issue is very much alive in recent literature on agricultural change and development, I want to explain my position at greater length here. The tentative arguments and conclusions presented below are separate from the body of the book. The reader who is experienced in economics or economic anthropology will quickly see that the various elements of my position are common ones, and that the differences with standard positions come in the combination and emphasis I have chosen.[1]

The Issue: Are Peasants Economic Men?

Wharton's summary at the conclusion of a major conference volume, *Subsistence Agriculture and Economic Development*, includes an excellent general statement of the issue:

[1] Although I may appear to be a formalist in the body of this book, economic anthropologists will see something of a substantivist in these pages. The "formalist-substantivist controversy" and the question of whether peasants are economic men are logically related issues. I have previously tried to show that the first is a bogus issue (Cancian 1966). And hopefully, the present argument will show that the second should join it among the empty arguments that impede both research and practical action.

Throughout the meeting a fundamental issue repeatedly raised concerned the dominance of economic versus noneconomic forces upon the economic behavior of subsistence farmers. For some participants the general conditions of subsistence agriculture automatically delimit an area where, on net balance, the noneconomic frequently outweighs the purely economic, leading to behavior that goes against the postulated behavior of economics. For others, the economic forces dominate the noneconomic, and the observed behavior patterns are considered quite consistent with the postulates of economics. (Wharton 1969: 456.)

This problem led to a controversy about the usefulness of economic theory for the study of subsistence agriculture. Wharton formulates three positions from the arguments of the many economists and other social scientists attending the conference. The extremes (around an intermediate position) are:

. . . those who accept the noneconomic dominance of economic behavior and therefore argue for a total recasting of economic theory to handle the economics of subsistence agriculture . . . [and] those who maintain that . . . subsistence and peasant farmers are highly rational and economic in their behavior, surmounting all such negative forces whenever the economic gains and returns outweigh the losses and costs. According to this view, leisure, work, thrift, and wealth with an eye to the marginal calculus are significant and are identical with, or not too dissimilar from, that which can be observed in modern societies. (1969: 457–58.)

Behrman, in his study *Supply Response in Underdeveloped Agriculture*, has also found a tripartite division useful in characterizing positions on a less general version of the fundamental issue:

The various *a priori* hypotheses about the supply responsiveness of underdeveloped agriculture to price changes may be divided into three major categories: (1) The hypothesis that farmers in underdeveloped agriculture respond quickly, normally, and efficiently to relative price changes. (2) The hypothesis that the marketed production of subsistence farmers is inversely related to price. (3) The hypothesis that institutional constraints are so limiting that any price response is insignificant. (1968: 3.)

T. W. Schultz and W. O. Jones (1960) were the early spokesmen for those who believe that economic theory is applicable to traditional economies. In his now classic *Transforming Traditional Agriculture* Schultz states: "There are comparatively few significant inefficiencies in the allocation of the factors of production in traditional agriculture" (1964: 37). Yotopoulos (1967) and Behrman (1968) have supported Schultz's general position in the context of detailed empirical studies of Greek (Yotopoulos) and Thai (Behrman) agriculture.

The classic exponent of the belief that economic theory is not appropriate for characterizing subsistence agriculture is, perhaps, J. H. Boeke. In his *Economics and Economic Policy of Dual Societies as Exemplified by Indonesia,* he writes:

Anyone expecting Western reactions will meet with frequent surprises. When the price of coconuts is high, the chances are that less of the commodities will be offered for sale; when wages are raised, the manager of the estate risks that less work will be done. . . . This inverse elasticity of supply should be noted as one of the essential differences between Western and Eastern economies. (1953: 40.)

Finally, two statements from the most recent edition of Samuelson's *Economics*:

In impoverished India, cows are sacred animals and, numbering millions, are allowed to walk through the streets foraging for food. While a naïve economist might regard these herds as a prime source of protein supplements to an already inadequate diet, the more profound scholar will take the psychology of custom into account when analyzing Indian economic development. (1970: 5.) . . . the Kwakiutl Indians consider it desirable not to accumulate wealth but to give it away in the *potlach*— a roisterous celebration. This deviation from acquisitive behavior will not surprise anthropologists; from their studies they know that what is correct behavior in one culture is often the greatest crime in another. (1970: 16.)

In sum, economists recognize a real difference between noneconomic or institutional factors and economic ones; and there is a real controversy between those who see the noneconomic (institutional) factors as dominant among subsistence agriculturalists and those who do not.

The Problem: There Are No Economic Men

Although students of agricultural development differ in the relative importance they attribute to economic and noneconomic factors in the economic life of subsistence agriculturalists, or peasants, they are virtually uniform in their implicit assertion that the comparison of economic and noneconomic factors is a valid one. I will argue that it is not— that economic man always operates within a cultural framework logically prior to his existence as economic man, and that this cultural framework defines the values in terms of which he economizes. This platitudinous restatement of the idea that the "given" institutional framework of an economic system may vary can be transformed into the conclusion that there are no true economic men, i.e. that there are no men whose economic activities are unaffected by their culture. If this is so, then perhaps the difference between men who respond to

"economic" incentives and those who apparently do not is a difference
in the degree to which observers have succeeded in specifying the institutional framework.

This section elaborates and defends these assertions in three ways.
First, I will try to suggest how the argument applies to the characterization of Zinacanteco economic life. Then I will briefly discuss our
own (urban Western) customs. Finally, I will try to characterize the
logical structure of the economic man model more fully.

Zinacanteco economic life. As I see it, the data in the body of this
book clearly support the position of those who see subsistence agriculture as dominated by economic forces and describable in terms of economic theory. The study's emphasis on the uncertainty characteristic of
a change situation suggests limits on the applicability of economic theory; but wherever information is adequate, and sometimes before information is adequate, Zinacantecos respond to economic incentives.
They have been presented as fully economic men hampered, as any
economic man would be, by lack of information.

However, as pointed out in Chapter 1, this picture of economic man
in Zinacantan is in substantial part the result of a deliberate effort on
my part, and it is not the only possible picture of Zinacanteco economic
life. I have tried to render all institutional constraints exogenous to the
economic system so that the Zinacantecos' economic decisions could be
analyzed in a way that would point up the importance of uncertainty
to their behavior in the face of change. But it is also possible to characterize Zinacantan in such a way that Zinacantecos appear tradition-
bound in their economic behavior. This approach may seem awkward
and artificial to the reader who has struggled through my description
of the organization of production in Chapter 3; but an emphasis on the
apparently unproductive aspects of Zinacanteco customs could easily
have been made to seem appropriate and natural at that time.

Imagine the following descriptions of the costs that are described as
recruitment expenses in Chapter 3.

Despite the fact that they must often sell their previous year's crop at
inferior prices in order to pay workers for weeding the new fields in
June, Zinacanteco employers usually pay for and drink liquor with each
worker when they recruit him in San Cristóbal, often to the point of
drunkenness. And when they return from the lowlands they provide a
feast including so much liquor that workers typically find it advisable to
sleep at the employer's house and return to their families the next day.

Fearful of being alone in the lowlands, Zinacantecos often risk damage
to their crops from weed competition by waiting for friends who have
decided to leave for the lowlands on a later day.

Though the fare to San Cristóbal is high, farmers often take only small amounts of corn to the market at one time. This practice permits them to enjoy a number of days in the city.

Clearly, these customs, and others that could be described in similar terms, represent situations in which Zinacanteco economic behavior is less than maximally efficient. I have shifted my emphasis to the institutional constraints, the customs, within which Zinacantecos work, but all economies have such an institutional setting.

Relativism and cross-cultural comparison. Though extreme to an anthropologist sensitive to the ultimate similarities of values across cultures, Samuelson's statement, "What is correct behavior in one culture is often the greatest crime in another," expresses the essence of relativism. I want to emphasize the aspect of relativism summed up in the statement, "We have customs too." Everybody knows this, but it seems to me that it is forgotten from time to time in the study of subsistence agriculturalists. Implicit (and occasionally explicit) in the application of the economic man model to subsistence agriculturalists is a comparison with modern economic systems, in which that model is presumed to describe a substantial part of economic behavior. Given our own customs and the logical nature of the model, this comparison is a silly one.

Our customs do not hinder our economic efficiency—not because we have unique customs, but because we define efficiency in the context of those customs. If we consider North American productive output in relation to North American customs, it is easy to see how production could be vastly increased by a simple alteration of custom. An obvious example is the day of rest, or weekend. Think of the capital now invested in plants, equipment, and church buildings that could be saved if we spread work and worship evenly over the days of the week, using factories and churches to their full capacity. The reduction in capital tied up in plants would probably not reach the 29 percent implied by adding two days each week to the five already worked, nor would the cost of churches be reduced by the 86 percent implied by eliminating six out of seven of them; but there is no doubt that the savings, and the increase in production effected by alternative uses of the capital, would be substantial. It is of course ridiculous to suggest that we completely abandon the social advantages of a common day of rest; but it is equally ridiculous to see ourselves as constantly maximizing economic goals.

In the first instance, then, there is no difference between our customs and those of others. The fact that we have managed to specify institutional constraints on our economic system so that we appear to maximize economic goals when engaged in "economic" activities is not a difference between our society and other societies. Rather, it reflects the vary-

ing degrees of experience we have in specifying the institutional constraints on our own and on other economies. Many analysts of modern
industrial society properly emphasize the degree to which economic
activities are differentiated from other aspects of life in such societies,
and properly contrast them with other societies in which economic life
is not as differentiated from other aspects of life. It is the Western
ideal—within very important limits—to divorce our productive activity
from many other aspects of life. However, this real difference between
industrial men and subsistence agriculturalists does not mean that the
former lead completely differentiated economic lives and the latter
wholly undifferentiated economic lives. The difference is one of degree.
From some perspectives it is a very important difference; and from other
perspectives it is a very small difference, given the logically conceivable
range.

Although some interpretations of the theory of pure capitalism suggest that our segregation of economic life may be complete, it clearly
is not. Like our tendency to behave according to the economic man
model, our isolation of economic life from other aspects of life is com-
promised by custom. We believe that workers should be hired for their
ability to work, not because their cousin happens to be the foreman;
and that college students should register on IBM cards because this
practice reduces the cost of training them for productive lives. On the
other hand, a foreman who fired a worker for a costly mistake might
have the sympathy of others under normal circumstances but would be
inviting censure if the mistake was made the week after the worker's son
committed suicide or if the worker, after twenty years on the job, was
beginning to lose his sight. Likewise, a university that requires students
to stand in line overnight in order to turn in cards they have previously
filled out will be criticized—not because the students will catch colds
and reduce their efficiency or because they have something more 'productive to do, but because the waiting is clearly not worth the savings
that might be effected. Today, in fact, even the limited degree to which
our economic life is differentiated from our other concerns is subjected
to increasing criticism by members of our own society.

In sum, we have customs too; and these customs quite appropriately
come before our total devotion to the abstract ideal represented by economic man. Although it is important to compare customs across societies, the question implied by comparing the relative influence of cultural
"economic" factors on economic decisions in different societies involves
very serious logical problems.

The logical structure of the hypothesis of economic maximization. My
principal assertion is that economic man always lives in a cultural con-

text, and that economic and noneconomic (traditional or institutional) forces for change are a noncomparable pair of features in any society. To elaborate this argument in its abstract or logical form, it may be helpful to begin with a distinction between the economizing, or maximizing, framework for viewing human behavior and the hypothesis of economic maximization. The first is a general framework in which human behavior is viewed as the outcome of decisions arrived at by the maximization of whatever goals need be attributed to the actor in order to make his behavior appear as maximization. The second, which embodies the economic man model in actual research, is an empirically testable hypothesis asserting that in many situations involving economic life men will allocate their efforts so as to maximize economic return. The first is a useful general strategy for organizing the study of an entire culture; the second is always tested within an existing cultural context.

Many sociologists and anthropologists would want to claim that they make independent measures of values and then use these to formulate a contingent hypothesis including both the economic and the noneconomic factors in decision-making, whereas economists include only economic factors. But I think sociologists and anthropologists often tend to use economizing or maximizing notions as a general framework that is not meant to be contingent in any way—that is, simply as a useful means of ordering their observations. In these terms, a successful study is one in which all values, attitudes, motives, and other impetuses to behavior are specified so that the actor appears to be maximizing (Cancian 1966; Burling 1962). By contrast, an economist who limits the goals of his actors to the maximization of economic returns is able to state an empirically testable hypothesis. He can ask: Given the following conditions, does the behavior in question produce a maximization of economic returns? And he can genuinely determine his answer to the question on the basis of empirical data.

However, in the controversy about the relative influence of economic and noneconomic forces on the economic behavior of subsistence farmers an economist is no better off with his limited hypothesis of economic maximization than are social scientists who use a general maximization orientation. This is so because economic and noneconomic (or better, "institutional") forces cannot be regarded as parallel elements in the operationalization of the theory represented by the hypothesis of economic maximization. No matter how the hypothesis of economic maximization is stated, it is always set within a huge *ceteris paribus* assumption about the institutional setting. Even in the most purely "economic" situations there are always a number of institutional constraints (often implicit ones). Thus whenever the hypothesis of economic maximization does not accurately describe human behavior, an economist has two options: he can accept the institutional constraints and assert the lack of economic maximization given his interpretation of these constraints; or he can change his picture of institutional constraints to cover those aspects of observed behavior that do not represent economic maximization. In this sense, the hypothesis of economic maximization is like any other hypothesis: failure to confirm it is never final.[2]

In sum, I conclude that because economic factors are always dependent on the prior definition of noneconomic factors it is impossible to state that noneconomic factors are more or less important than economic factors. As a result, comparative research that asserts the dominance of economic factors in one place and noneconomic factors in another is meaningless.

Internal Comparison and Internal Differentiation

If institutional constraints (i.e., noneconomic factors) can be taken as constant and unvarying, then the problem of their logical priority is circumvented and comparative research is possible. Here, I want to describe two types of research that meet this condition and are potentially useful to the policy maker who hopes to aid the economic development of a society or group. Both internal comparison and internal differentiation research focus on comparative rates of change within a group and avoid comparisons across groups.

The internal comparison approach seeks to directly answer the interesting question of what kind of program will be most effective for a particular group. This approach is developed and well illustrated for

---

[2] Within the framework of normative economics, the finding that the hypothesis of economic maximization does not apply to a given situation has important practical consequences. These are situations where the economist's client sets the constraints and is presumably prepared to live with the consequences—where the economic-man framework and the hypothesis of economic maximization are used to test allocative efficiency given available resources and other constraints. Within the framework of positive, or descriptive, economics, where the goal is simply a description of behavior, not optimal allocation given known constraints, the tables are turned, and defining the constraints becomes the crucial empirical question. Although the efficient-allocation problem of normative economics and the institutional-definition problem of descriptive economics do not offer commensurable elements, it is clear that in many situations changes in constraints provide massive power over outcomes when compared with changes in allocative efficiency like those sought by the economist. Concrete examples of this balance between institutional constraints and allocative problems were presented above.

economists by Matthew Edel in a paper entitled "Innovative supply: a weak point in economic development theory" (1970).[3] He says:

The study of innovated supply has also tended to ignore a more important contribution to development policy implicit in the dichotomy of motivations it keeps rediscovering. It is not that any group of people is either all-responsive to the market or not responsive at all to anything but direct promotion of mental or social change. Rather, in most cases, either market stimuli or direct promotion may do something to increase effort in new lines of production. The most economical mix of policies to achieve a given production of a new product may include a combination of both *subsidies* to price (to induce more output in proportion to the elasticity of response) and direct *promotion* (through business motivation courses, community development programs, advertisement of the attraction of urban life, or other means). (Edel 1970: 18.)

In discussing his own comparative study of the effectiveness of promotion and subsidy in the Colombian Acción Comunal program Edel makes the point that is fundamental to the internal comparison approach: knowledge of the comparative effectiveness of different programs (subsidy and promotion in his case) can be used to determine the most effective mix for fostering development within the society studied.[4]

The internal differentiation approach used in Chapter 8 of this book illustrates the second type of research relevant to the administration of resources intended to produce economic development. Rather than differentiating programs, as Edel has, I have differentiated types of individuals into those more and less apt to take the chance typically necessary for an innovator. This approach, or any other application of a general theory to identify the segments of a population most apt to innovate and thereby get change moving, does not determine whether the group in question responds quickly or slowly in comparison with other groups that might receive the same allocation of resources. Rather, like internal comparison, it identifies a way to most effectively introduce

[3] Edel is interested in the treatment in development theory of businessmen and factory workers as well as peasants, and finds parallel arguments for each.

[4] Another part of Edel's argument is his stress on the importance of learning in the process of change. "Interactions of promotion and subsidy are made more complex by the learning that can take place from participating in new activity. Both individual familiarity with an activity, or the observation of the success of others at it, can improve the willingness of people to undertake it, or their effectiveness in it." (1970: 18.) And, "The debate between 'traditionalism' and 'elasticity' theories has ignored learning effects as well as intermediate degrees of elasticity." (1970: 23.)

economic change once the decision to commit resources to a particular society or group has been made.

Cross-Cultural Comparison, Policy, and the Ethics of Aiding Development

Cross-cultural comparisons of economic responsiveness were criticized above because the institutional constraints that must be defined in order to make them are not constant and unvarying across cultures; hence the outcomes of the comparisons vary with the definition of the constraints. The further conclusion that such comparisons cannot be made at all is untenable from a practical point of view: in the course of human affairs, men must and will make comparisons and evaluations.

Here, I want to briefly explore the policy implications of the argument I have made about the logical structure of research on economic responsiveness. My argument leads to ultimate despair only if we seek in research some objective, scientific, and impersonal means of making comparisons and evaluations. If we are willing to accept the setting of institutional constraints as a fundamentally human, nonscientific act we have no research problem, for the logic of research described above permits comparisons after the institutional constraints are defined. We are only required to remove the cloak of scientific respectability from the process of defining constraints.

Once this is done, the policy maker is restricted to a more limited set of questions for research. If we frame his questions in terms of economic aid between nations, he may ask two kinds. First, *given* a commitment of so much aid to a particular nation that has a *given* set of goals, how can this aid be used most effectively? This is the question of internal comparison discussed immediately above.[5] Or he may ask: *given* these resources and *given* this program, where will it produce the greatest material return to the dollar?[6] However, he may not ask who are economic men and who are tradition-bound men. Nor may he justify his allocation of resources on the basis of an answer to this question.

[5] In cross-cultural comparisons we are setting uniform constraints across known and recognized diversity; above, we assumed that uniform constraints were appropriate to the entire domain within which the comparison was made. But if cross-cultural comparison involves the potentially arbitrary setting of institutional constraints across cultures, then any comparison faces similar dangers across subcultures, across social and economic ranks, and ultimately, in the infinite regress, across individuals. Only societal consensus and an assertion of humanistic values can rid us of this problem.

[6] Of course, the setting of goals (e.g., achieving maximum total material production or Pareto optimality) must be recognized as a value-determining act that necessarily has no objective rationale.

Since the first two questions seem sufficient to sustain the efforts of nations to promote economic development, the limitation placed on the third question may seem trivial and irrelevant in practice. I do not think it is. Although it might be trivial if we lived in a world or a nation that carefully moderated the value it placed on economic achievement, we clearly do not live in such a world or nation. As long as men value economic men, the spurious attribution or denial of that quality to other men will make an important difference and must be done in a guarded manner that denies it ultimate significance.

Ideas about peasant irrationality have virtually passed out of the scientific vocabulary and have become exclusively a part of the public and political domain; to almost all scientists interested in agricultural development, all men are now rational. But the practical problems that originally brought on the notion of irrationality have not disappeared, and the old labels have been replaced among scientists by the more subtle distinction between those who respond to economic variables and those who are tradition-bound. Frank recognition of this fact may save us from another round of creative labeling and help us advance research where it can be useful, i.e., within a clearly acknowledged cultural-moral framework that is logically prior to the research. The alternative is to hide morality and culture in a spuriously objective science, thereby dehumanizing our morality and confusing our science.

Bibliography

ANDSA
  1964 Esquema social y economico de los estados de la republica Mexicana, II (Campeche, Coahuila, Colima, Chiapas). Mexico, D.F.

ATKINSON, JOHN W., and NORMAN T. FEATHER
  1966 A theory of achievement motivation. New York, Wiley.

BAHR, ADELAIDE PIRROTA
  1961 The economic role of women in San Felipe. Unpublished ms., Harvard Chiapas Project.

BEHRMAN, JERE R.
  1968 Supply response in underdeveloped agriculture: A case study of four major annual crops in Thailand, 1937–63. Amsterdam, North-Holland.

BOEKE, J. H.
  1953 Economics and economic policy of dual societies as exemplified by Indonesia. New York, Institute of Pacific Relations.

BUNNIN, NICHOLAS F.
  1966 La industria de las flores en Zinacantan. In Los Zinacantecos, Evon Z. Vogt, ed. Mexico, Instituto Nacional Indigenista, Colección de Antropología Social 7.

BURLING, ROBBINS
  1962 Maximization theories and the study of economic anthropology. American Anthropologist 64: 802–21.

CANCIAN, FRANCESCA M.
  n.d. Measuring norms and their relation to behavior in a Maya community. Forthcoming.

CANCIAN, FRANK

1965a Economics and prestige in a Maya community: The religious cargo system in Zinacantan. Stanford, Stanford University Press.

1965b Efectos de los programas económicas del gobierno mexicano en las tierras altas Mayas de Zinacantan. *Estudios de Cultura Maya* 5: 281–97.

1966 Maximization as norm, strategy, and theory: A comment on programmatic statements in economic anthropology. *American Anthropologist* 68: 465–70.

1967 Stratification and risk-taking: A theory tested on agricultural innovation. *American Sociological Review* 32: 912–27.

CAPRIATA, JORGE

1965 Economic and social behavior of Zinacanteco corn middlemen. Unpublished ms., Harvard Chiapas Project.

COLLIER, GEORGE A.

1968 Land inheritance and land use in a modern Maya community. Unpublished dissertation, Department of Social Relations, Harvard.

1969 Computer processing of genealogies and analysis of settlement pattern. *Human Mosaic* 3: 133–41.

COLLIER, JANE F.

1968 Courtship and marriage in Zinacantan, Chiapas, Mexico. Middle American Research Institute Publication 25: 139–201. New Orleans, Tulane University.

1970 Zinacanteco law: A study of conflict in a modern Maya community. Tulane dissertation, to be published in book form by Stanford University Press.

CONKLIN, HAROLD C.

1961 The study of shifting cultivation. *Current Anthropology* 2: 27–61.

DEAN, ALFRED, HERBERT A. AURBACH, and C. PAUL MARSH

1958 Some factors related to rationality in decision-making among farm operators. *Rural Sociology* 23: 121–35.

DITTES, JAMES E., and HAROLD H. KELLEY

1956 Effects of different conditions of acceptance upon conformity to group norms. *Journal of Abnormal and Social Psychology* 53: 100–107.

DORFMAN, ROBERT

1964 The price system. Englewood Cliffs, N.J., Prentice-Hall.

EDEL, MATTHEW
  1966 El ejido de Zinacantan. *In* Los Zinacantecos, Evon Z. Vogt, ed. Mexico, Instituto Nacional Indigenista, Colección de Antropología Social 7.
  1970 Innovative supply: A weak point in economic development theory. *Social Science Information* 9: 9–40.

EDWARDS, WARD, and AMOS TVERSKY, eds.
  1967 Decision-making: Selected readings. Baltimore, Penguin Books.

EL UNIVERSAL
  1964 La producción de maíz subió casi 100%. *El Universal*, June 20, 1964, p. 9.

FLIEGEL, FREDERICK C.
  1957 Farm income and the adoption of farm practices. *Rural Sociology* 22: 159–62.

GROSS, NEAL C.
  1942 The diffusion of a culture trait in two Iowa townships. Unpublished thesis, Iowa State University, Ames.

HELBIG, KARL M.
  1964 La cuenca superior del Río Grijalva: Un estudio regional de Chiapas, sureste de Mexico. Tuxtla Gutiérrez, Instituto de Ciencias y Artes de Chiapas.

HILL, A. DAVID
  1964 The changing landscape of a Mexican municipio: Villa las Rosas, Chiapas. Chicago, University of Chicago, Department of Geography, Research Paper No. 91.

HOMANS, GEORGE CASPER
  1961 Social behavior: Its elementary forms. New York: Harcourt, Brace, and World.

JONES, WILLIAM O.
  1960 Economic man in Africa. Stanford, *Food Research Institute Studies*, I (2): 107–33.

KNIGHT, FRANK H.
  1921 Risk, uncertainty, and profit. New York, Kelley.

LINDSTROM, DAVID E.
  1958 Diffusion of agricultural and home economics practices in a Japanese rural community. *Rural Sociology* 23: 171–83.

MARSH, C. PAUL, and A. LEE COLEMAN
  1955 The relation of farmer characteristics to the adoption of recommended farm practices. *Rural Sociology* 20: 289–96.

McClelland, David C.
  1961 The achieving society. Princeton, N.J., Van Nostrand.
Plan Chiapas
  1962 Plan Chiapas. Secretaria de Agricultura y Ganaderia. A pamphlet with no further citation information.
Plattner, Stuart M.
  1969 Peddlers, pigs, and profits: A community of itinerant peddlers in southeastern Mexico. Unpublished dissertation, Department of Anthropology, Stanford University.
Pozas, Ricardo
  1959 Chamula: Un pueblo indio de los altos de Chiapas. Mexico, D.F., Memorias del Instituto Nacional Indigenista 8.
Price, Richard S.
  1968 Land use in a Maya community. *International Archives of Ethnography* 51:1–19.
Rogers, Everett M.
  1962 Diffusion of innovations. New York, Free Press.
Samuelson, Paul A.
  1970 Economics. New York, McGraw-Hill.
Schultz, Theodore W.
  1964 Transforming traditional agriculture. New Haven, Yale University Press.
Simon, Herbert A.
  1957 Models of man. New York, Wiley.
Stauder, Jack
  1966 Algunos aspectos de la agricultura zinacanteca en tierra caliente. *In* Los Zinacantecos, Evon Z. Vogt, ed. Mexico, Instituto Nacional Indigenista, Colección de Antropología Social 7.
Vogt, Evon Z.
  1969 Zinacantan: A Maya community in the highlands of Chiapas. Cambridge, Mass., Harvard University Press.
  1970 The Zinacantecos of Mexico: A modern Maya way of life. New York: Holt, Rinehart, and Winston.
Wharton, Clifton R., Jr.
  1968 Risk, uncertainty, and the subsistence farmer: Technological innovation and resistance to change in the context of survival. New York, Agricultural Development Council. Mimeo.
Wharton, Clifton R., Jr., Ed.
  1969 Subsistence agriculture and economic development. Chicago, Aldine.

WILKENING, EUGENE A.
1952 Acceptance of improved farm practices in three coastal plain counties. North Carolina Agricultural Experiment Station, Bulletin 98.

WILKENING, EUGENE A., JOHN GARTELL, and HARTLEY PRESSER
1969 Stratification and innovative behavior: A reexamination of Cancian's curvilinear hypothesis. Paper presented at the annual meeting of the Rural Sociological Society, August 1969, San Francisco, Calif.

YOTOPOULOS, PAN A.
1967 Allocative efficiency in economic development: A cross section analysis of Epirus farming. Athens, Center for Planning and Economic Research, Research Monograph Series 18.

ZUBIN, DAVID A.
1963 The San Cristóbal corn market: A discussion of vendor-buyer interaction. Unpublished ms., Harvard Chiapas Project.

Index